CW00517875

Rural Pest Control

Steve Caple

Rural Pest Control

COUNTRYMAN PEST CONTROL LTD

Countryman Pest Control Ltd
25 Eldean Road
Duston
Northampton NN5 6RF
Email: steve@countrymanpestcontrol.co.uk

ISBN 10: 0-9553896-0-7
ISBN 13: 978-0-9553896-0-3

Cover image © Steve Caple

Typeset in 11pt Bembo by Troubador Publishing Ltd, Leicester, UK
Printed in the UK by The Cromwell Press Ltd, Trowbridge, Wilts, UK

In memory of my dear friend Mr Albert (Bert) Williams of Little Brington with whom I spent so many happy hours talking all things rural.

Contents

Acknowledgements

With grateful thanks to the editors of **The Countryman's Weekly** and **Today's Technician** for kind permission to reprint some of the writings in this book. Thanks also to Mr M Munns for his help in preparing the finished manuscript. Extracts from the **Defra code of practice on the use of snares in fox and rabbit control** are © Crown copyright.

1
Introduction

'It's a long road that never ends', someone once said to me. I've always remembered it, and in a way, it sums up the apprenticeship of country life that most rural pest controllers have to serve before having sufficient knowledge to embark on a career as a professional hunter. I was born in a lovely little Northamptonshire village called New Duston in the early 1950s. I still live in the same village today, just a stones throw from the house I grew up in. The village has changed completely from the one I remember as a boy. The fields and meadows that gave so much enjoyment to the embryo hunter have long since disappeared under a swathe of tarmac roads and housing developments. The old overgrown ornamental lake where we used to fish for goldfish using a bent pin on a length of cotton tied to a hazel wand has now been buried deep below ground, underneath a cul-de-sac of neat semi-detached houses. If they only knew what memories lay buried beneath their neatly trimmed lawns and colourful borders? The countless childhood hours spent arm deep in the murky waters, trying to catch newts and frogs to cram into our jam jars, the moorhen's eggs we used to collect to cook on the camp fire, and the masses of frog and toad spawn we used to gather to take home by the bucketful, are all now a long distant memory.

Although I didn't realise it at the time, those halcyon days spent wandering the fields and hedgerows as a boy were going to instil within me a love and respect for the countryside and all things in it, which endures to this day. I firmly believe that you need to have spent your formative years learning the ways of the animals and birds and the field craft that goes with it, to really understand the role of a rural pest controller. By this I mean, unless you have spent time studying the habits of those creatures you are asked to control, you will find it extremely difficult to work out a successful strategy to ensure their removal. That said these things can be learned by anybody who is prepared to put in the long hours of hard work that is needed to become proficient at the job. This book will show you the way I tackle certain situations and problems, but what it can't teach anyone is the 'feel' for the job. This only comes with years of practical experience. If I have whetted your appetite, and you are new to rural pest control, many of the methods and scenarios described in this book will give you some tips and ideas that should help you along the path of your chosen profession. If you are an old hand at the game, you may find that most of what is written here you already know, but we can all learn something new at times. Indeed, in rural pest control you never stop learning. There is always some new little trick or ruse that we can use to our advantage, so if you find something within the pages of this book that improves your success rate with a particular species then it will have done its job. As with all things appertaining to the countryside, nature is never an exact science, because what works one day may fail abysmally the next. It doesn't mean you have done anything wrong (though you may have), it simply means that on that particular day Mother Nature didn't play ball. I have tried to keep the text as simple and as uncomplicated as I possibly can, because that is the way most pest control operations are carried out. Remember, the day you become over-confident is the day you start to fail. Sadly, we live in an age where country pursuits are being hounded mercilessly by the anti-anything brigade. We owe it to those who went before us to keep these traditions alive. Future generations must be able to enjoy the same freedom and pleasure of the countryside as we do today, so to that end, I make no apologies for producing this manual with the hope that it may help keep these traditional methods alive.

2
Reasons for Control

When does an animal, bird or insect become a pest? Well, usually when it comes into conflict with man and instils fear, causes damage, or spreads disease and contamination. In an ever-increasing rural encroachment, vast swathes of beautiful countryside are systematically being turned into sprawling urban developments. Housing estates, fast-food outlets, cinemas, warehouses and retail parks have now become common place on the outskirts of most major towns and cities. It is little wonder then that the natural inhabitants of such areas are sometimes forced to compete for survival in what was once **their** territory. Take the mole for example, one minute he is tunnelling freely in open pastureland, next he is causing damage to someone's newly laid lawn. Many times I have been called to a new housing development to remove a mole that has 'suddenly appeared overnight'. My explanation that the mole was there long before they were usually falls on deaf ears. In an established rural environment the mole is tolerated to a certain degree and it's only when the numbers increase to such a level that valuable grazing or silage is in danger of becoming contaminated, that action sometimes becomes necessary.

A particular species of animal or insect only becomes a pest because of where it is and not what it is. A wasp nest in an old rabbit burrow located in some remote woodland is not a problem; the same nest on a school playing field would rightly be regarded as a major hazard. Rabbits located on vast areas of moorland miles from the nearest arable farm are not really going to do a lot of damage, but the same rabbits undermining a railway embankment, or consuming large acreages of growing crops, becomes a serious pest that definitely needs controlling. Grey squirrels in parks and gardens naturally give pleasure to many people, but when the same people who have been feeding them in the garden find they have taken up residence in their roof, they tend not to be quite so enamoured with them. If shown the damage to the electrical wiring in a roof space caused by a family of squirrels, most people would never encourage squirrels into the garden again.

It is a fact of life that some animals have to be controlled. Some may disagree and often do, and I have had many a heated argument with well intentioned but sadly misinformed people who are vehemently opposed to pest control in all its forms. For instance, when confronted by people who are against the killing of foxes or rabbits, I always ask them what they would do if they had an infestation of rats in their home. The usual response is that they would get someone in to get rid of them, and from that moment in time they have lost the argument.

Not all control methods involve the killing of animals or insects, because in many cases proofing or relocation can be the answer. I have lost count of the number of bumble-bee nests I have transported to a safer environment and the bee swarms I have re-united with grateful owners. On feral pigeon jobs, proofing an area is always preferred over the culling of the birds (culling is only ever carried out as a last resort). The only drawback with pigeon proofing is that it tends to push the problem somewhere else. Often reassurance and advice is all that is needed to resolve certain pest problems. Every Christmas I get numerous calls to remove wasp nests from roof spaces, the nests having been discovered when the decorations are required for the tree. After explaining the nest is empty and all the wasps are dead, the problem is quickly resolved.

To summarise then, insects, birds or animals only become a pest and need controlling to prevent contamination or disease, economic losses, or to avoid damage and distress. Unless a particular creature falls into the above categories, then the best solution is to do nothing and leave well alone. Sometimes certain species, bats or badgers for example, do qualify for the term 'pest' under the nuisance or fear factor, but these are protected by law and mustn't be interfered with. On the subject of law, there are many legislative acts that must be adhered to when carrying out pest control operations. The sensible pest controller will familiarise himself with all the relevant documentation before attempting to control any species of bird or animal. The professional pest controller should always carry out his duties safely and humanely and stay within the remit of the law at all times.

3
Rabbit

Before embarking on a career as a professional pest controller I'd had a few different jobs. I had been a labourer on a building site, a glazier, and had trained for three years as a psychiatric nurse in a local hospital. I had reached a point in my life where my job as a glazing manager had become tedious and was looking for a change of direction to put some inspiration back into my mundane lifestyle. Fate stepped in to save me when it was announced that the company I worked for was relocating to Oxford and our branch was closing down. For many years I had been helping local farmers, landowners and householders with their pest problems, so this seemed as good a chance as any, to turn a much loved hobby into a full-time occupation. Grabbing the meagre redundancy payment with both hands, I took the decision to throw caution to the wind and try to establish a small rural based pest control business. The money I received enabled me to buy a battered old 4×4, a few cages and traps, and paid for an advert in a national advertising directory. I had missed the deadline for the advert and mine wasn't going to appear until the next year's addition, so I had to find another way to advertise my fledgling business. I decided to write to all the local golf clubs in the area asking if they had a rabbit problem and offering my services. One golf club took up the offer, a price was agreed and a year's contract drawn up; the rest, as they say, is history. I had a regular monthly income, the mortgage could be paid and the business was up and running.

Professional rabbit control is not the same as doing it for sport or as a hobby. The methods may be exactly the same, but the only difference is that, *every* rabbit, no matter how big or small, has to be removed. This often involves the unsavoury but necessary task of killing pregnant or doe rabbits with young, something that many people may find distasteful. If you are only controlling rabbits for sport, you can obviously be more selective. So let's take a look in detail at the many and varied methods of rabbit control available to us. There's ferreting, shooting, snaring, cage trapping, long netting, lamping with gun or dog, drop box and spring trapping, and gassing using aluminium phosphide to name just a few. Each method has its own merits, but some methods are only suitable in certain situations.

When looking at any new job it pays to spend a little time doing a thorough survey of the surrounding area. A client may well have rabbits under a garden shed, but if there is open farmland to the rear of the garden, then this is where the problem needs addressing. To remove the rabbits from under the shed is easy; what's more difficult is preventing re-colonisation from the adjoining farmland. It may be that rabbit-proof fencing needs fitting, a costly and time consuming exercise. But where do you draw the line? If you net the bottom of the garden, do you also have to net the sides? You see, nothing is straightforward in rabbit control. Every situation is different in some way and needs to be judged on its merits. Let's take a look at the many options available to us in more detail.

Ferreting

Ferreting has got to be the professional rabbit catcher's most useful weapon in his armoury of control methods. Ninety percent of all rabbit control will make use of the humble ferret; please note I use the word control, rather than eradication (which is seldom achievable). In fact, never ever tell a potential customer that you will get rid of all the rabbits; always say you will remove as many as you possibly can.

So what do you need to start ferreting? Well, if you are just ferreting for sport you need only a ferret and a few purse nets. If you intend doing it for a living, you will need between four and twelve ferrets and over 200 purse nets. Add to that four or five 50 yd long nets, ditch nets, hedge nets, gate nets, etc. and you begin to see that full time rabbit control can sometimes become a mammoth undertaking. Of course, not all jobs require this amount of equipment. Some, like the rabbit under the shed, need just half a dozen purse nets and a ferret to resolve. On some larger contracts you will use every ferret and net that you own. You need to be prepared for any situation you're called to deal with if you choose to undertake professional pest control.

There's no need to go into great detail on the husbandry of ferrets since there are many books that have been written on the subject already. All I will say is, if you keep ferrets, treat them with respect; keep their living quarters clean, warm and dry, and handle them as much as possible. It always amazes me that, after the ferrets have been put away at the end of a long hard season, some owners never handle them until they begin ferreting again. A well-cared for ferret will give years of loyal service if you treat it properly. I prefer to use Jills on my ferreting jobs. All are fitted with transmitter collars, and I would never work a ferret without one. That said, I always make sure I have a large Hob ferret with me because sometimes he will shift stubborn rabbits where the Jills have failed.

Take care in selecting purse nets for rabbit jobs, some nets work better than others in certain situations. If I'm working in woods containing a lot of bramble and thorns, I prefer to use heavy nylon nets. These tend not to tangle very easily and any debris caught in the mesh can quickly be shaken out. If, on the other hand, I'm working sandy banks or open grass buries, then hemp or waxed cotton nets are more suitable. I tend to make all my own nets these days, replacing old and worn nets as and when they become unserviceable. Some nets currently in use are over 30 years old and still in excellent condition. The secret is making sure they are never left in the net bag in a damp condition. Always hang nets up to dry after every ferreting trip since they will last years longer if they are allowed to dry naturally after becoming wet. The beauty of making your own nets is that you can choose the size and type of material used in their construction. The shop-bought nylon nets are usually 2 inch mesh. I make mine an eighth bigger than this on a standard purse net. I also make purse nets with a 1 inch mesh that I use when ferreting buries that are known to contain small rabbits. The only problem with using these small nets is that ferrets tend to pull them off the holes when trying to get through the mesh. Much has been said in the past regarding the colour of nets. With purse nets I think it makes little difference what colour they are. I'm of the opinion that when a rabbit comes from a dark hole into bright light, the net colour becomes irrelevant. I have used blue, black, white, green, brown and orange nets in the past, and have never noticed any reluctance on the part of the rabbit to bolt into one. With long nets though, I think colour does make a difference. I have seen rabbits turn away from light-coloured nets, especially at night, even when travelling at speed. I have never seen a rabbit turn away from a green or brown long net at night (or in daylight for that matter).

Let me give you an example of the type of contract a professional rabbit man may be asked to undertake as part of his daily workload. Late one February I was asked to price a job for a client who had suffered severe damage to a 3.8 hectare wood. The wood was relatively young, having only been planted some 20 years previously with native British species. The adjoining farmland had been left 'set a-side' for a season and the rabbits were struggling to find decent food. Subsequently, they had begun to strip and ring the bark of the young trees up as far as they could reach. The ash trees in particular had come in for special attention. I had never before seen ring barking to such an extent and it was clear to see that, unless drastic action was taken immediately, the entire wood was in danger of being killed off.

A thorough inspection of the wood was undertaken and it was clear this was going to be a far from straightforward operation. To start with, the wood had been planted on top of an old landfill site. This was patently obvious because of the amount of bottles and plastic containers that were strewn about, having been unearthed by the rabbits. The buries were very shallow and the soil quite light, which realistically

Tip, the Bedlington × Whippet type hybrid watching over the nets

Rabbits can do a lot of damage to trees in immature woodland

meant gassing using aluminium phosphide would not be a satisfactory solution and certainly not cost effective. The only sensible option would be to systematically ferret each bury and warren within the wood. With some of the warrens having over 100 holes it was going to be a slow and laborious undertaking.

A price was agreed and a pretty hefty one at that. I knew I had several days' hard work in front of me and that it was not going to be easy. Many of the warrens and buries were covered in dense bramble, but fortunately the client was able to arrange contractors to clear the undergrowth from above the holes prior to work commencing. This was to prove a problem later on though; the chopped pieces of briar were to play havoc with the nets.

Day one of the operation arrived, it was a glorious frosty morning, perfect ferreting weather in fact. I had two experienced ferreters to assist me as this was one job that couldn't be done single handed. The plan of attack was this; we would start at the largest warren and net every hole, we would then net all the holes on the surrounding smaller buries in the hope that any rabbits slipping the nets would be back netted as they tried to get back to the main warren. We had eight ferrets with us and my 8-month-old lurcher pup, a Whippet x Bedlington type hybrid (but more of him later). My original plan had been to enclose the warrens with long nets as a back up but the shredded pieces of briar would soon have rendered them utterly useless.

A couple of inexperienced ferrets were entered first as it's always advisable to let them have first crack of the whip while the rabbits are plentiful. They worked well. Soon the rabbits started moving, bolting at nice regular intervals, giving us just enough time to re-position fresh nets before the next arrival. We must have had about ten rabbits in as many minutes before things started to slow up. It was time to enter the experienced workers. Things were on the move again and the reluctant bolters were soon shifted by the seasoned campaigners. As yet, we hadn't had a rabbit refuse to bolt, which was remarkable really, since you usually get one or two that decide to sit tight. We had 20 rabbits in the bag when we decided all that was likely to come out had come out and it was time to move on. All the nets were left in place while we ferreted the surrounding buries, a decision that was well justified as a few rabbits that slipped the nets, were back netted as they tried to return to the main warren. The briars were now beginning to become a problem when a few rabbits were able to throw the nets off and make good their escape. The surrounding buries were soon cleared of rabbits and we now had a grand total of 32; things were going well. Now came the least enjoyable bit. All the nets now had to be picked up (we had over 200 out) and the holes filled in. What! I hear you cry, the holes filled in. Well that's the difference between professional rabbit control and ferreting purely for sport. You see, any rabbits that slip the nets will head back for the safety of the burrows. If these are left open they are lost. By filling the holes, the rabbits are forced into buries yet to be ferreted.

Well satisfied with the day so far, we adjourned for coffee and a bite to eat whilst deciding on the next plan of action. We agreed to finish the day doing some smaller buries, leaving the larger ones until another day; we had at least another three days work ahead of us. The day progressed at a steady pace, with a constant supply of rabbits from the smaller buries, even some single holes producing up to four rabbits.

The highlight of the morning for me though, was when a rabbit slipped the net and my lurcher pup set off in hot pursuit. The rabbit made for the big warren where we had just filled in the holes; the dog picked it up as it hesitated momentarily outside a blocked hole. Only an experienced dog handler would know how I felt at that moment in time, watching him proudly retrieve the rabbit alive back to hand. I don't mind admitting I had a lump in my throat because you see, until that morning, he had never seen a live rabbit.

We finished the first day with a total of 46 rabbits. Not a bad start, but we had missed many due to the nets getting snagged and we hoped the missed rabbits would be in the warrens we had yet to ferret. The second day went much the same as the first when we did one big warren and finished the day doing the smaller ones. The big warren did not produce the number of rabbits we had expected and we only finished the day with 26, which was a bit disappointing. The second day was memorable for one remarkable achievement though. I had a cameraman filming a ferreting sequence for a future DVD production and the camera was set up to film a 15-hole bury. A couple of rabbits bolted and were caught both in the net and on film. Another rabbit bolted before I could get a net back on and the pup set off in hot pursuit. Another rabbit bolted and was quickly dispatched, when I noticed the cameraman was no longer filming me. He was filming the dog coming back through the trees, and yes, he was carrying a live rabbit in his mouth. You couldn't have planned it if you'd tried. As stated previously, Tip, the lurcher pup, is a Whippet × Bedlington type hybrid, a blue-black rough-coated dog from Phil Haynes the renowned lurcher man. I told Phil at the outset that I wanted a dog to work with ferrets and nets; I didn't need a leggy speed merchant as I'd already got one of those ready to train on the lamp. He hasn't disappointed because I can honestly say he's the most intelligent dog I have ever owned. By the age of six months he had over a dozen partridge to his credit, all retrieved live to hand. He had learned to catch them under the hedge against the rabbit netting, something I had never taught or encouraged him to do. All the birds were released totally unharmed (minus a few feathers perhaps) to fly another day. Anyway, back to the wood, the third day was very slow and we struggled for rabbits. Most of the big buries had been done and we were concentrating on the very small ones, but we still finished the day with another 15 rabbits, not a lot, but 15 less to damage the trees.

I estimated we had one more day left to complete the job. The most difficult buries, the ones straddling the perimeter fence, had been left until last. The weather forecast didn't look good, heavy rain was predicted and for once it was accurate. When we arrived it was bucketing down but we had all brought a change of clothes and decided to work through the rain to get the job finished. Once again

You only need a ferret and a few purse nets if you are just ferreting for sport

we started at the largest bury and netted over 60 holes. We were to be disappointed as it only produced four rabbits much to our annoyance. The reason for the lack of rabbits I put down to the day before when I saw a large polecat or feral ferret working along the ditch we were now working in. I think he must have cleared most of the rabbits overnight. We were by now getting soaked to the skin but we still had time to laugh at one of the young ferrets dancing outside one of the holes. She had never encountered rain before and was snapping at the rain drops falling on her back. It was becoming very hard work, the mud in the ditch was getting worse, and the nets were sodden, as indeed we were, but another light-hearted moment cheered us up. A rabbit cleared a hole, the result of a double bolt, and once again the pup set off in hot pursuit, only this time the rabbit decided to make its dash for freedom across a waterlogged field of bean stubble. It was so wet and muddy that both the dog and rabbit were running in slow motion. We all cheered the dog on until he eventually caught the rabbit at the fence, by which time they were both at walking pace. This was his third catch in three days. We ended the day with another 17 rabbits, not as good as we'd hoped for but quite reasonable considering the atrocious conditions. We changed out of our wet clothes for the journey home and my thoughts at last turned towards a nice hot bath. We had taken a total of 104 rabbits from the wood. Many were missed but that's the nature of ferreting, you can't catch them all.

That then, was the type of situation you may have to deal with if you choose to undertake rabbit clearance as a profession. It can be very hard work at times and if the client is paying you handsomely for your efforts, you must produce results. Well, if you want to work for them again you do!

Cage Trapping

Another successful method of rabbit control is cage trapping. When used in the right location it can prove very productive. My cage traps are used mainly on small garden jobs and some spectacular results can be achieved at times. One of the more frequent requests I get during the spring and summer months is to remove rabbits from under sheds in gardens where if left to breed, can cause immense damage to bedding plants and vegetable gardens. The rabbit under the shed is a fairly straightforward process. A few purse nets strategically placed, a long net wrapped around as a back up, slip in the ferret, and a few thumps and bumps later, one rabbit in the bag. In an ideal world all jobs should be as simple as that but experience has taught me that life isn't always quite as easy. If the call came in early summer then invariably the rabbit under the shed would be a doe with anything up to seven youngsters. This may rule out the use of the ferret if the young are very small. We now have a problem; we could sit out for a few nights with an air rifle in the hope of picking off a few young ones, but this would not be a cost-effective option for the client. This is a typical situation where the use of carefully positioned cage traps would provide the ideal solution.

So, how do you go about it? Place the cages around the shed, bait with carrots, and wait for the client to give you a ring when you have a catch. It sounds simple enough, but there are other things that can be done to increase your chance of success. The first things to look at are the cages themselves. The cages I use are 230 mm wide, 760 mm long and 260 mm high. I have used smaller cages but have found that rabbits are more reluctant to enter them. Before using any trap the first thing you must do is 'weather' it. The cage would have passed through many hands before arriving on your doorstep and you will need to remove all odours acquired during manufacture before it is ready for use. To obtain that weathered look, I routinely dip all my new cages in a water butt and scrub them with a stiff hand brush prior to standing them outside for a few weeks. Whilst on the subject of scents, don't make the mistake of storing your rabbit cages in the shed next to the cage you have just caught a mink in. You won't catch a rabbit for months if you do. This applies to all cages. I keep my squirrel cages solely for squirrels and my mink cages only for mink. I never mix the two. If you don't clean your cages between catches they take on the scent of the target species which often proves irresistible to the next visitor to your trap.

So now we have the cages all clean and ready to set. The next step is to find the best location to obtain maximum results. Most people, unless really experienced in trapping rabbits, tend to do one of two things when first setting cages. They put them directly outside rabbit holes or they put them where a rabbit comes through a wire fence, hoping in both cases that the rabbit will run straight into the cage. Although you will occasionally catch rabbits like that, there are far more successful places to set cages that will greatly increase your catch rate.

Let us return briefly to our rabbits under the shed. If you look carefully around the area you should detect a line of rabbit runs leading away from the shed and out to a favourite feeding area. It's somewhere along this run or runs (there may be several), that you want to set your cages. For those who don't know what constitutes a rabbit run, these are the little patches of flattened grass, like stepping stones, that the rabbit uses on his nightly travels. The cages want setting not on the runs, but to the side of them, so the rabbit has to pass the entrance to your cage on his way out to feed. The further out from the shed you can set the cages the greater your chance of success. Several cages can be placed along the same run but these need to be at least 3 m apart and on alternate sides of the run for best results.

The most popular bait by far has to be carrots. I have tried lettuce, parsnip, and cabbage among others, but have always had the most success with carrots. These should be sliced into quarters long ways, for the best results. I have had results in the past using cornflakes or rabbit food from the pet shop, but these tend to be rendered ineffective if it rains overnight. Even if rain isn't a problem, it tends to attract birds and rodents, so for me, it's carrots every time. Some people recommend sprinkling a little salt on the carrots to enhance the attraction (rabbits are supposed to like salt) but personally I haven't noticed any difference. All that remains is to set the cage to the side of the run, put a few carrots in the back and entrance of the cage, and set the treadle plate. The cage is now ready to catch its first visitor. One very important point is to make sure the cage sits firmly on the ground and doesn't move about. Push a stick through the cage if necessary to hold it tight to the ground. A rabbit will not enter a cage if it feels it moving,

Don't be put off if the cage doesn't catch for a day or so. This is quite normal because the rabbits need to get used to the cages before they gain enough confidence to enter them. Once they start catching, another useful tip is to expel the urine from a recently dispatched rabbit into the cage bottom; this makes the trap far more attractive and inviting to the next passing rabbit. You don't need to cover or disguise the cages in any way because the rabbits will enter them quite freely. You, or the client, must check the cages at least once a day (more often if you are expecting a heavy catch, especially during hot weather), and

You can find many things other than rabbits caught in cages

Carefully positioned cages will catch rabbits of all sizes if given a chance

the occupants humanely dispatched. I check my cages at first light and again at dusk since we have a duty to ensure that any trapped animal does not suffer unnecessary stress. The cages can look slightly odd sitting way out in a field beside the rabbit runs, but they will catch plenty of rabbits if you give them a chance. If you are setting cages on land that others have access to, it pays to set them at last light and pick them up again first thing in the morning. This minimises the chance of someone walking off with, or damaging, your expensive cages. It is very common to find cages have been rolled over and moved quite a distance from where they were set. This is nearly always a fox or a badger trying to get at the rabbit inside the cage. If this happens regularly you will have to pin the cages to the ground. I have had the local blacksmith make me some stainless steel T bars about 18 inches long that I push through the cages to hold them tight to the floor.

As an example of how effective cage trapping can be, I was called in by an old customer of mine who was being plagued by rabbits coming into his garden from a large wood at the rear of his property. Six cages were placed just inside the garden on the runs that came through the hedge. In one week I removed 16 rabbits of varying sizes. After three weeks I had trapped over 40, including a large white tame rabbit that had absconded from a garden further down the road.

On my golf club contract, along with shooting and ferreting, cage trapping forms a major part of my control programme. I run a total of 30 cages at any one time, moving them around to different parts of the course to problem areas as the need arises. Most years I remove between 1,500 and 2,000 rabbits from the golf course, so they really do work.

Don't let me give the impression that cages catch at all times because sometimes I get it completely wrong and catch no rabbits at all. This is usually down to me setting them in the wrong places. I don't always get it right first time. As with all things, experiment a little with locations; if you are not catching in one area, moving the cages can often pay dividends. It would be fair to say that if the cages haven't caught for over a week then they should be moved to somewhere different. If you can find a fence that rabbits run along, these can be deadly catching places. Here I like to place cages end to end, so which ever way a rabbit comes along the fence it finds the inviting entrance to a cage. Make sure the cages aren't touching each other though, otherwise a trapped rabbit will trigger the other cage. Take care when setting cage traps along fences since many other creatures like to run the fence lines at night. I once thrust my hand into a cage at dusk to remove what I thought was a baby rabbit, only to find it was a large rat curled up in the corner. I have caught many things in rabbit cage traps over the years. Foxes and badgers are most common but I've also caught mink, ferrets, cats, hares, partridge and pheasants. The list is endless, so always be careful when checking cages.

One final piece of advice. When removing a rabbit from a cage, always stand the cage on its end and reach in and get it. Don't do as I once did and lift the door up and reach in with the cage still on the ground. In my case the result was one escaped rabbit and one rather sheepish pest controller left holding an empty cage.

Lamping with Dogs

Lamping with dogs is not really a method to use for professional rabbit control. Used for a couple of nights to pick up outlying rabbits after a ferreting trip lamping can be quite productive, but for the majority of rabbit jobs I get asked to look at, it would not be a sensible option. There are far more successful and easier ways of taking rabbits. As a sport though it's hard to beat. I love to see a good dog run the beam and retrieve rabbits live to hand and that's why I'm training my lurcher bitch on the lamp so I can go out for a couple of hours at night, to help unwind after a hard days work (talk about a busman's holiday). As I said, it's not really a method I would recommend for wholesale rabbit clearance but, used in the right situation and as part of a control programme, it can have its merits.

Snaring

If ferreting is the most useful method for rabbit control then snaring comes a close second. On large rabbit clearance jobs (if conditions are suitable), a weeks snaring prior to ferreting can greatly reduce the rabbit numbers. I have on occasions, after heavily snaring an area, found the surrounding buries virtually devoid of rabbits when ferreting them later. In October 2005, DEFRA launched a new code of practice on the use of snares. The **Defra code of practice on the use of snares in fox and rabbit control** was the result of a complete review of the use of snares in England and Wales. Among the recommendations contained within the code are the fitting of permanent stops to snares and reviewing the times the snares are inspected. The code is not a statutory document but one which lays down best practice. At the time of writing it is still perfectly legal to set snares for rabbits and foxes and, as long as the guidelines contained within the code are adhered to, the use of snares will be seen to be effective without having a detrimental effect on other wildlife.

With all trapping methods the trap is only as good as the person setting it. I can show anyone in a few minutes how to set a snare, Fenn trap, or cage trap, but what I can't teach anyone is the 'feel' for the job. This only comes with years of experience out in the field. Being able to look at a rabbit run and know instinctively which pad to set the snare on, gauging how fast the rabbit is likely to be travelling, or how high he is carrying his head, all these things take years to learn. It isn't called field-craft for nothing! Most people new to the art of snaring think the large pads where the rabbit rests and leaves plenty of droppings is the place to set the snare. This is completely wrong. The snare is set on the smallest of pads between two slightly larger ones, where the rabbit only momentarily lands before jumping on to the next pad. It is during this leap between the two pads that the rabbit is snared.

Let's start at the very beginning and examine the rabbit runs themselves. These little patches of flattened grass, like stepping stones, are what the rabbit uses on nightly travels to and from favourite feeding areas. They are very distinctive and easy to see close to where the rabbits come out from the buries, becoming less obvious the further out into the field they travel. Somewhere along this line of pads will be the ideal place to carefully position a snare, to intercept the rabbit on night-time manoeuvres. For anybody new to snaring I always recommend they buy or make no more than 20 or 30 snares to begin with. Setting snares takes the beginner a long time. Looking for the right run, adjusting the setting height, or checking that the tealer isn't across the run, all increase the setting time for the novice trapper. The snares themselves want to be the eight-strand type with eyelets; nowadays most shop-bought snares are of this type. I always replace the tethering cords of the shop-bought snares with strong nylon cord; as this tends to adds a few more months to their working life. Next, an ash holding peg is needed to fix the nylon tethering cord. Ash is well suited to the wet and dry conditions encountered during the working life of a snare, and can take a lot of hammering into the ground before needing replacement. The nylon tether wants to be about 12 inches of doubled cord. If you then tie a knot between the snare and the holding peg to form an elongated figure of eight, you can slip your trapping hammer through the loop to pull the snare out of the ground when you come to lift the snares. The last thing you need is a tealer. This is the little stick that holds the noose of the snare above the rabbit run. They are called different names depending on which part of the country you come from. They are also called a pricker, or a tiller, the latter derived from the fact that you 'till' a snare, as in setting. No doubt there are other names out there for it, but for our purposes we will stick with tealer. The tealer is usually made from straight hazel cut into 7 inch (18 cm) lengths and needs to be about 10 mm thick. They might need to be cut a little longer if snaring on light or sandy soil. Next a roof-shaped cut needs to be made at the top end, with a half inch split down the length of the wood, to hold the snare wire in place. The bottom end wants sharpening, not to a point, but to a chisel-like shape, as this helps stop the tealer turning around when set beside the run.

The snares now want hanging outside for a month or so to 'weather' and lose the newness and shine acquired during manufacture. If you are doing a lot of snaring you can always have replacement wires hanging outside ready to replace any lost or damaged snares. The snares need to have a stop crimped on them about 5 inches (14 cm) from the 'eye' to conform to the new DEFRA code of practice, and want fixing on now. You can short-cut the weathering process by soaking the wires in a strong solution of tea for a few hours, or you can dull them down by dipping them in boiling water containing a little soda. Walnut husks and oak bark will also darken snares, the choice is yours. I prefer to hang the wires outside and let the weather do the job for me. In reality, the snares don't have to be black to catch successfully. If they are a dull golden colour they will still blend in with the bleached grass tufts found on most meadowland. They will eventually turn very dark with use anyway. Some people advocate the use of gloves when setting snares or traps but I never wear them personally; I rub my hands in a convenient mole hill if there's one about to help mask any scents, but other than that I don't bother. I've never noticed any difference in my catch rate by not wearing gloves. The best way to carry snares is to tie them in bundles of ten and to pass the holding pegs through the snare loops, which helps keep the snare noose open. Right, we now have our snares ready to go; you have already chosen a likely field, you have checked with the landowner that there will be no livestock in the field for a few days and also that no one will be walking the dog or horse riding across the field, the weather forecast is favourable, dark and windy with no rain expected; and you are full of anticipation and ready to start.

When arriving at any new snaring ground I find it beneficial to walk up and down the field a few times to ascertain the extent of the rabbit problem. First I walk slowly along the fence line or hedgerow making mental notes as I go as to the number of well-used rabbit runs leading out into the field. I then walk back along the fence line but this time about 15 m out into the field, again counting the runs as I cross them. I then walk the field a third time, maybe 30 m out, again noting the

The night's catch from 40 snares having been set along a woodland fence

Snaring rabbits is an art anyone can learn given enough practise

number of rabbit runs. I now have a pretty good idea how many snares I need and how far out in the field to set them. As a rule, the runs become less distinct the further out from the buries they go. I like to set my snares as far out as possible before the runs start to disappear, this may be as little as 5 or 10 m from the fence line, or as far out as 40 m. It all depends on the strength of the runs and the amount of rabbit activity.

I once had to control the rabbits in a long narrow horse paddock; the horses were only going to be away for a few days, so I had to act fast. Shooting was not an option due to the tenant's agreement, and ferreting was impossible due to the impenetrable mass of brambles that covered the hedgerows. Snaring was the logical answer to the problem since the rabbits were in such numbers that they were eating more grass than the horses. It was a snaring paradise. The field was only 50 m wide and about 400 m long and the rabbit runs went in straight lines from one side of the field to the other, like the lines on a shove-halfpenny board. I walked down one side of the paddock about 10 m out from the hedge, setting a snare on every run as I crossed it. I then walked back up the other side of the paddock, again setting a snare on every run I passed. I put out precisely 100 snares, 50 each side of the paddock; in theory I had two snares set on each run. Such were the rabbit numbers I had caught six before I finished setting the last snare. It was nearly dark by the time I got back to the motor and I had just finished loading the gear into the back when a squeal pierced the rapidly fading light, then another, and another. Taking the torch, I retraced my steps back down the paddock. There were rabbits running everywhere; some were held in snares and some ran into the wires as I approached. I had never seen such rabbit activity before. I took a further 14 rabbits from the snares and reset them before I finally left the field. Next morning at first light I checked the snares and removed 26 rabbits; a fox had also removed four or five. I did another check at midday and had caught another nine and a further check at dusk accounted for another 16. The numbers then continued to decline until, by the third day, I was taking only four or five rabbits at a time

Rabbits can cause a lot of damage to growing crops unless they are controlled

Rabbit runs leading from a wood out onto pasture land

from the snares. It was time to bring a halt to proceedings. In three days snaring I had removed over 100 rabbits from the paddock, the fox had his share too, and the client was delighted to see the rabbit population greatly reduced and his valuable grazing saved.

Anyway, I digress, we have arrived at the field with our snares and we have made sure they are weathered and haven't been stored next to the ferret box in the motor (some people do, believe me). Let's assume we have found 50 good runs leading from a railway embankment, through sheep netting and out into the field. We have decided we will set the snares about 15 m out from the fence where the runs are quite distinct. What we need to look for is a pad about the size of a beer mat or smaller, certainly not bigger, and if it's between two pads of similar size, then this is where we set the snare. The reason you set on the small pad is because the rabbit only lands there for an instant before hopping to the next pad; it's during this leap to the next pad that he's snared. Now we have decided where we need to set the snare, we can insert the snare wire into the tight slit at the top of the tealer, and shape the noose to the desired size. There are differing opinions on the size of a snare noose. I like mine slightly on the large size while others prefer them smaller. It all comes down to the individual and how he sees it at the time, this is where the 'feel' for the job comes in. As a general rule the noose wants to be about 5 inches (125 mm) long and 4 inches (100 mm) deep and roughly pear shaped about the size of a clenched fist. With the wire held firmly in the tealer the snare now needs positioning directly above the centre of the chosen pad and the tealer pushed into the ground. The bottom of the snare loop wants to sit about 4 inches (100 mm) off the ground or higher if you are snaring in longer grass. As a good guide, it wants to be a hand's width off the pad but this must be guessed since grass tends to hold human scent for quite a while. Make sure the tealer is not overhanging the run. Stand back and look at it as a rabbit would approach it. Once you are satisfied with the setting you can hammer the holding peg into the ground. The snare should be held firmly within the split of the tealer and the eyelet should rest on the top. The noose wants to be fashioned in such a way that it wants to spring open. If it is correctly set, you should be able to gently brush the palm of your hand against it and have it spring back into place. Snares have to withstand very windy conditions at times, so a secure setting is vital.

That's it; you proceed to set the remaining snares on the runs in the same manner, keeping them in as straight a line as possible to make life easier when the time comes to lift them again. If you carry the snares in bundles of ten, tie the cord that is holding the bundle together onto the cord of the first snare you set. Repeat this routine with each bundle of ten. When the time comes to lift the snares, if you get to the second cord and you only have nine snares, you only have a little way to go back to find the missing wire. If you don't do this and get to the end of the field and find you have only 46 snares, you have the entire length of the field to search for the missing snares; which is not a nice prospect when you are wet, muddy and tired. I like to leave the snares down for three nights before moving them to another location; most of the

A rabbit snare correctly set above the pad on a well used run

rabbits will have been caught if you have done your job properly, unless the weather has been particularly bad. If it has been wet or there has been a period of hard frost, snares can be left on the ground for longer, but as a rule, move them after a few nights otherwise the rabbits will become snare shy. Don't set snares on that particular field again for at least two weeks or longer if possible. It needs to rest to let the rabbit numbers build up again and to let them forget about the wires.

Foxes and badgers are a constant problem where snares are set; they hear a rabbit squealing and are over to investigate in an instant. They soon learn that a free meal awaits them, conveniently pegged down to the ground. There is not an awful lot you can do to prevent it, other than checking the snares frequently and removing the rabbits. Some people advocate replacing the snare cord with light chain. The rattling is supposed to deter foxes; personally, I think it would scare more rabbits from using the runs rather than deter foxes and badgers. If when checking snares you find just a rabbit's head in the wire, you can be

Fifty rabbit snares and spare tealers ready for setting

certain that a fox is responsible; increase your inspection times if needs be, because one thing is certain, the fox will be back! Snares need checking often anyway and I visit mine at first light, midday, and again at dusk. Often a visit an hour or so after dark can be beneficial to remove trapped rabbits. Don't think every snare you set will catch a rabbit; if 1 in 3 catches you will have done well, often it is less than this. If I set 100 snares I would expect to pick up between 30 and 40 rabbits in a night depending on the amount of rabbit activity in the area. I have had as few as 12 and as many as 70 in a night from 100 snares. It all depends on many things but the weather plays a major role in determining rabbit activity as does the fact there may be a fox, badger or human prowling about. In very heavy dew, for instance, rabbits won't travel very far at all.

So there you have it, snaring is an art that can be learned by anyone prepared to put in the hours of work needed to become proficient at the job. If you are unsure of your ability, see if you can find someone to show you the ropes before you start, then, as you become more confident, try setting a couple of snares yourself until you get the hang of it. I have described here the most basic method of setting a snare. No two people set a snare exactly the same. I have told how I do it and what works for me. Some people never set on the pad, preferring to set on the jump (between the pads), which obviously works for them. You must find the way that works best for you and perfect it. If we are to keep this valuable addition of rabbit control in our armoury then we must be seen to comply with the guidelines set out in the DEFRA code. If we don't then the anti-field sport lobby will be only too quick to clamour for an all out ban on snares.

Long Netting

Traditionally the long net has always been seen as the poacher's supreme weapon; dropped silently at night it has accounted for many thousands of rabbits that would otherwise have remained out of reach to all but the privileged few. Before the days of myxomatosis, large hauls could be taken on a regular basis. It was quite usual for 100 plus rabbits to be taken nightly, although to achieve this, more than one drop of the nets was usually necessary. Weather conditions were paramount when deciding to go on a night-time excursion and the really clever long netters would only venture forth when conditions were conducive to a large catch. Dark windy nights with a hint of rain were preferred. The wind direction was the most important factor to be taken into consideration when deciding where to drop the nets; also, no moon had to be visible in the night sky. A targeted field may well have been heaving with rabbits sitting way out from the buries, but, if the wind was blowing in the wrong direction it would prove a fruitless task deploying the nets.

I have, over the years, been on many night time excursions with the long nets but have never achieved the large bags supposedly caught by others. Until recently, the most rabbits I'd ever caught in one drop of the long nets was 18. That particular night the conditions were strongly in my favour with a stiff breeze blowing from the feeding rabbits down below a ridge in the field, which meant the setting of the net went undetected at the hedgerow. But just recently I surpassed that when I took 28 from a secluded football pitch in two 50 yd long nets. The head lines of the nets were virtually on the floor for its whole length such was the weight of the rabbits tangled in the mesh. I have taken over 80 rabbits in a single night, but this entailed visiting three different locations and several drops of the nets to achieve. I work using four 50 yd traditional long nets and one 75 yd long net on the quick set system as devised by that well known nets man Brian (Bryn) Brinded. The quick set system consists of a long net with the holding pegs permanently fixed to the head and foot lines. The net is deployed from a specially designed carrying basket which enables the lone operator to run the net out at walking pace. It can be picked back up again just as quickly. At night I find 50 yd nets much easier to work with than the longer nets. The 100 yd nets can become very heavy after a time, especially if wet.

To be successful with the long nets at night requires much practice by the active participants. Nets should be run out and picked back up again many times during daylight hours so each member of the netting team (usually three) knows exactly what they are expected to do under the cover of darkness. When everything becomes second nature during the day, it is then time to venture forth on a suitable night to put those skills to the test.

I'll tell how we used to do it; it worked for us many times. There is no right or wrong way to drop long nets provided certain rules are strictly adhered to. So-called experts, and we all know what the definition of an expert is (an ex is a has-been and a spurt is a drip under pressure), tell us it must be done this or that way to be successful. Rubbish! If it works for you and it catches rabbits then that is the right way. We worked with four 50 yd nets and about 40 setting pegs, one person carrying the nets draped over their arm with the pegs in a quiver over their shoulder. The second person took the first net, pushed the anchor pin into the ground, and walked along backwards letting a hank of net slip off the pin a few feet at a time. I walked after them setting the net up on the pegs that were being handed silently to me from the quiver at roughly 5 yd intervals. This was repeated with the other three nets. Not a word was spoken during the entire procedure and, in a matter of five minutes or so, the nets were up, linked, and ready to catch. One person stayed with the nets while we went the long way round to the far end of the field and began walking back towards the hedgerow where the nets were set. We could now talk as much as we liked. The human voice is great at getting rabbits to move and this, combined with the slapping of a couple of spare long net pegs against our Wellingtons, sent the rabbits fleeing back home towards the waiting nets. Any rabbits that hit and tangled in the nets were 'chinned' by the net man and left lying white belly uppermost for collection later. I mentioned certain rules that must be adhered to when long netting at night, they are: (1) Rabbits must be in sufficient quantities to justify dropping the nets; (2) Weather conditions must be perfect, dark with the wind in the right direction; (3) Setting of the nets must be done in absolute silence and remain undetected by the rabbits. Satisfying these three basic requirements should bring a successful outcome to the night's exertions whatever method you use to drop the nets. A lot of nonsense has been written in the past about long netting at night; some of these armchair experts would, I think, prove sadly lacking if put to the test on a real night out. Setting a long net after dark is a lot more difficult than doing it in daylight. Any debris picked up in the net at night will prove to be the very devil to remove from the mesh. For this reason, it can be beneficial to walk the ground due to be netted in daylight and, any twigs, thorns or briars removed prior to setting the nets later that evening.

As well as the traditional long net there are other types of long net systems available such as the drop net and the trammel net. The shop-bought nets are mostly made of 4 oz or 6 oz nylon and usually have a 2 inch mesh. I like my nets an eighth bigger than this because they hold rabbits better. With the slightly bigger mesh the rabbits can force their heads through the netting but are unable to withdraw them because they get trapped by the ears. The trammel net is a twin or triple walled net; one side has small mesh whilst the other is of much larger mesh. The idea is that the fleeing rabbit hits the small mesh and takes it through the larger mesh where it twists and hangs in a pocket rather like a purse net. The drop net is really a system whereby the long net (usually a trammel net) is hoisted up on a row of sliding poles that allow rabbits to pass under the net and out to feed. The whole net can then be dropped to the ground by the pest controller, by pulling on a remote cord. This system is useful where rabbits are unapproachable due to over lamping or where street lighting makes a traditional setting impossible.

As a pest control method, most long nets see more use in the daytime than at night. They are an essential part of rabbit control equipment, especially when used in conjunction with tricky ferreting situations. Imagine if you will, a blackthorn hedgerow, with sheep netting either side, with an impenetrable mass of bramble to negotiate, impossible to get purse nets on, and you have been contracted to clear the rabbits.

You can catch a lot of rabbits using the long net in daytime

The night's catch from one drop of the long net

This is the perfect scenario for the deployment of long nets; two nets run out down the length of the hedge each side, as far out as possible, linked to another couple run though or at the end of the hedge. Having the nets out as far as possible enables rabbits to get well clear of the warren before encountering the meshes of the long net. This also tends to stop them trying to get back to the perceived safety of the warren which they would attempt to do if the nets were set nearer the hedgerow. One other advantage of having the nets set a long way out is to give the dogs a better chance to pick up any rabbits that may slip the nets and head back for home.

Even with the warrens surrounded by long nets, I still like to put purse nets on any bolt holes out in the field and where rabbits come through or under the wire fence. To carry out a large scale operation like this obviously requires the assistance of several people who understand what the job entails. The kind of helpers that are easily bored, or worse still, easily excitable, are to be avoided at all costs. It pays to position oneself as close to the hedge as possible and not to stand out in the field beside the nets. A rabbit will sit waiting to bolt until the ferret is almost upon it before finally making a dash for freedom. If it sees you standing way out in the field, it may decide to face the ferret rather than bolt towards the net. I like to place a person at each hedge net to intercept any rabbits that run up or down the hedge and a person each side to watch the field nets. Any rabbits that bolt are running away from the perceived danger area and should hit and tangle in the nets at full speed.

You don't have to set long nets very high off the ground. My setting pegs are only about 24 inches long and, when pushed into the ground, leave the net standing 18 to 20 inches high. I have never seen a rabbit attempt to jump a long net yet; they always try to push through them, so the most important thing is to have plenty of loose netting (bag or kill). If you have the nets too tight the rabbits bounce off and don't tangle in the mesh. Some nets have the kill tied in, as in the quick set system, but I prefer to have free running mesh on my traditional long nets. On the subject of long net pegs (I like mine cut from straight

The quick set long net system (75 yards) complete with carrying basket

The traditional long net set sees more use in daytime than at night

hazel), I always put a small split in the top of them to facilitate the top (head) line of the net. Now this is not strictly correct, the top line should be held tight by a couple of half-hitches in the traditional method of setting, but for sheer speed of use, I find the split peg a lot quicker and easier to use. Nowadays, the hazel setting peg has largely been replaced by fibreglass versions but, being a traditionalist at heart I'll stay with the wooden ones for now, if only for the fact they are easily replaced should one get lost or damaged.

Most of my daytime long-net work involves the removal of rabbits from awkward situations. Chicken runs, sheds, summer houses, woodpiles, you name it and I've probably caught rabbits there. If I'm asked to remove rabbits during the summer months, it will invariably be to remove a doe and her litter. Sentiment must play no part in professional rabbit control; all rabbits, whatever the size, must be caught and humanely dispatched. When ferreting under sheds, it pays to put a person at opposite diagonal corners. That way, two people can watch all four sides for bolting rabbits. Take care when ferreting under sheds though, especially chicken coops, because rats will very often bolt alongside rabbits. It's amazing how often the two are found together. When dealing with rabbits in long nets, it pays to extricate them as quickly as possible from the netting, for two reasons. First, the enmeshed rabbits tend to pull the head line down to the ground, allowing following rabbits to avoid the net. Second, a rabbit left in the net for any length of time (especially at night), will start to chew his way out of the net.

On a recent contract I was asked to remove rabbits from the gardens of a rehabilitation hospital. It was a brain injury unit and the scrapes in the lawns were becoming a danger to the patients as they strolled in the gardens. There were very few holes to ferret as the majority of rabbits lived above ground in the overgrown flower beds, so it was a simple matter to encircle the beds with long nets and let the ferrets wander freely through the undergrowth. We caught 36 rabbits from the garden in a morning. One incident in particular stood out on this assignment.

There was a single rabbit hole, hidden deep in the middle of a flower bed; a purse net was dropped over and the ferret slipped in, when almost immediately a rabbit exploded into the net, followed by another, then another, and another. In all, eight rabbits bolted from the hole and were caught in the nets. To this day, that remains the greatest number of rabbits I have ever bolted from a single hole.

Briefly returning to long netting at night. I always like to use brown or green long nets with dark coloured head and foot lines. If I owned light coloured hemp nets, I would certainly be looking to dye them a different colour. I can't see the point of going to all the trouble of silently dropping 200 yd of netting at night, only to have it stick out like a sore thumb to the approaching rabbits. This is just my personal opinion you understand. If you work light coloured nets and produce results, then who am I to argue? You must always use what works best for you.

Drop Box Traps

Drop box traps are very useful if you have a long term contract to control rabbits. Set on well used runs through rabbit proof netting, they can be deadly at catching rabbits that would otherwise remain outside the boundary fence. They can only be used where you have a long-term control programme because they are permanently set into the ground. The traps themselves are made of wood or galvanised metal (the latter being the longer lasting) and consist of a holding chamber (sunk into the ground) with a tunnel running across the top that contains a counterbalanced floor that tips the unsuspecting rabbit into the chamber below, before returning to the set position. Any suitable rabbit netting fence will do, though they are best installed as part of a fresh rabbit-proofing exercise. By that I mean, where rabbits have been regularly using an entry point, to a golf course for example, if the netting is installed across the run and the trap tunnel inserted through the wire, it will catch all rabbits that routinely use that particular way in. After installation, a drop box should be left in an unset position for

Rabbit drop box trap showing counterbalanced arms prior to installation

Galvanised metal rabbit box trap showing arms in the set position

40

as long as possible to allow rabbits to start using the tunnel with confidence. A heavy weight, such as a log, needs to lay across the counterbalance arms to keep the floor in position until the rabbits are regularly using the tunnel on a nightly basis. When the rabbit numbers using the run are deemed sufficient, the log can be removed, and the trap left in the catch position. The first night's catch is usually the biggest, and numbers will then gradually decline until it stops catching. It is then time to close down the trap until the rabbit traffic builds up again. When lifting the lid on the inspection chamber to view the night's catch, be aware there may be other occupants apart from rabbits waiting in the trap. Rats, hedgehogs, stoats and feral ferrets are quite common. Once I had a very irate tabby cat that wasn't overjoyed to see me after being shut in all night. As a precaution I have a wire mesh cover that sits under the inspection chamber lid so I can view the occupants before thrusting my hand into the abyss below. Any unwelcome visitors (provided they are vermin) can be humanely dispatched with a suitable air rifle.

The metal box traps are better than the wooden variety in as much as they don't warp and twist like wood is prone to do when wet. They are available from a number of sources and a quick look on the Internet will provide a list of suppliers. Suitably disguised with vegetation, the box trap becomes all but invisible to the unsuspecting eye of its intended victims and, when left to weather and blend in with the surrounding flora, soon becomes an accepted part of the landscape. Used as part of a regular rabbit control programme, drop boxes can prove extremely effective if sited in the correct location. A lot of work is required to install the trap system correctly, so it is essential the best location is selected first time. Some people run a trap line of boxes, placing them every 100 yd or so along rabbit netting fences. Obviously there would need to be considerable rabbit activity to warrant the work and expense involved but some fantastic catches can be achieved. Sometimes as many as 300 rabbits can be taken in a single night from a line of boxes, but as previously stated, there would have to be a substantial supply of rabbits to achieve this.

Shooting

Rabbit control using shotgun, air rifle, or .22 rimfire is another highly effective method of clearing large numbers of rabbits. At harvest time a few guns strategically placed around a field being combined, can make vast inroads into the local rabbit population. Many times I have been involved in such shoots, and have seen hundreds of rabbits taken out of a single field. Beans were always a favourite crop to find lots of rabbits in. With two other guns we once shot 76 rabbits from a tiny three acre field of beans. On the last cut by the combine the rabbits were shuffling down in droves in front of the cutter blades. We had to signal the driver to slow down because we feared that when he got to the end of his run, the rabbits would all bolt at once. As it was, many were lost because we couldn't reload the guns fast enough (the barrels of my shotgun were nearly too hot to hold). After this particular harvest year (1990) we finished up shooting 568 rabbits from five fields of wheat and beans, none of which were more than 10 acres in size. The surrounding hedgerow buries were nearly devoid of rabbits when we arrived to ferret them in the winter.

Shooting over ferrets can be another profitable exercise if used in the right situation. Here I'm referring to those places that are impossible to purse or long net, for example dense woodland or old quarry workings. In these situations, shots should only be taken when rabbits are well clear of the holes, so as to avoid potential injury to the ferret, and also to prevent wounded rabbits going back down into the bury. Guns, once positioned, should all know exactly where they are allowed to take a safe shot and shouldn't, under any circumstances, move from that position until the shooting is over. Most of my rabbit shooting is done at night using the .22 rimfire with a sound moderator fitted (sometimes erroneously called a silencer). I honestly don't know where I'd be without this versatile piece of equipment. It's the gun I take everywhere, zeroed in at 60 yd it accounts for thousands of rabbits every year. Most nights of the week, weather permitting, I drive around my rabbit contracts after dark, taking out any rabbits foolish

Harvest is a good time for shooting rabbits (76 from 3 acres of beans)

Shooting at night with the .22 rifle is another way of removing rabbits

enough to sit momentarily in the headlights. This constant nibbling away at the rabbits, night after night, keeps the numbers down to a manageable level. Some rabbits quickly get wise to the sound of the 4×4 and head for home the instant they hear the engine and see the lights. Others, especially the young of the year, hesitate at the hedgerow and pay the price for their curiosity. If snaring a field before a ferreting session is beneficial, then shooting the same field after a ferreting session can be even more so. Until the disturbance at the buries has been forgotten about, many rabbits will be very reluctant to venture back to the hedgerows. Human scents, ferret odours and dog smells linger for quite a while after a days ferreting. Some rabbits appear confused and can often be found sitting out at night; these tend to squat down rather than head for home, and can easily be dealt with using the .22 rifle and a lamp. Obviously, if you are using the .22 rifle at night, you must always be aware of potential dangers. You must know the ground you are shooting over like the back of your hand, and you must know where you can take a safe shot and where you can't. Does a footpath run across the land? Are there sheep or cattle behind the hedgerow? Could someone be walking the dog? All these things and more need to be taken into account before pulling the trigger. If in any doubt, don't fire. It's just not worth taking the risk for the sake of another rabbit.

The shooting of rabbits by a full-time pest controller does not rely solely on the shotgun and .22 rimfire. There are many situations where these would be too noisy or powerful for the job in hand. Some operations need to be carried out quietly and discreetly and, this is where the air rifle comes in. I use a .177 air rifle for all the jobs that require a 'silent' approach. I find it invaluable in many situations. Primarily I use it for feral pigeon control in roof spaces or warehouses, but it is just as useful when sitting out at dusk to mop up a family of young rabbits. The golf course I have a contract on is a good example. The course is dotted with ancient 500-year-old sweet chestnut trees, the bottoms of which are honeycombed with rabbit holes and, being warm and dry, are favourite breeding places. On warm summer evenings I position the 4×4 some 30 yd from a selected tree, out of range of wayward golf balls, and pick off the rabbits one by one as they emerge to feed. The streams of golfers that meander past the trees are blissfully unaware of the rabbit cull going on in their midst. Entire families of young rabbits can quickly and safely be eliminated using this method. I prefer the .177 over the .22 air rifle as I think it has more penetrative power; it is certainly powerful enough to cleanly kill an adult rabbit at 30 yd with a carefully placed head shot. The use of scopes on an air rifle is almost obligatory nowadays (although not essential). There are hundreds of variations out there, but all I will say is, choose the most expensive type you can afford, at a magnification that is best suited for the work you intend to do. Remember, the more powerful the magnification, the steadier you will have to hold the gun. I find the most suitable scope for general purpose vermin control is a 4–12 × 40 set at 6 × magnification. The beauty of the modern day pneumatic multi-shot air rifles is that they are virtually silent in use so, provided you remain still and in a concealed position, many shots can be taken before the intended quarry becomes aware of your location.

The diminutive .410 shotgun is the type most country boys first learned to shoot rabbits with, when out for a bit of sport with the ferrets on a Sunday morning. It is excellent for bowling rabbits over at ranges out to 25 yd, but much further than this, it starts to become ineffective, I know it will kill rabbits out as far as 35–40 yd at times, and I've done it myself, but, this is an exception rather than the rule. I don't use the .410 shotgun for pest control nowadays, preferring instead to use the more robust 12 bore; some people though, still find a use for this handy little gun at times, but, as previously stated, I think it is more a sporting weapon than a serious pest-control tool.

So, the choice is yours, always choose the weapon that is most suited for the job you are undertaking; it's common sense really. You wouldn't sit in a garden blasting away at rabbits under a shed with a 12 bore, any more than you would use an air rifle on a harvest rabbit shoot (or would you?).

Spring Trapping

Another productive method of rabbit control is the use of spring traps. In certain situations it can be the most successful control method of all. The most common spring trap in use at present has to be the Mk6 Fenn, followed by the Imbra and the Juby. The last two are no longer in production (as far as I'm aware) but, at the time of writing, are still legal to use. The one drawback to serious trapping is that at times it can be a very labour-intensive and time-consuming exercise. The Spring Trap Approval Order 1995 lists the types of traps that may legally be used and their conditions of use. All pest controllers should familiarise themselves with this order and the Pest Act 1954 (Section 8) before attempting to set any type of spring trap. Before moving on to the setting of spring traps, it might be interesting to go back a little in time, and look at the origins of rabbit trapping as a profession.

The old-time rabbit catchers, or warreners as they were known, used to make a good living trapping rabbits to sell in the meat markets of the large towns and cities throughout Britain. Vast quantities of rabbits were put on overnight trains and sent down to London from remote stations across the country. The light sandy soils of Norfolk (especially around the Thetford area) held large numbers of man-made warrens that produced a constant supply of fresh meat for the wealthy inhabitants of the cities. The rabbit skins were also much sought after. These were used in the hat and felt making trades and a good pelt would often fetch more than the meat itself. It was the need for undamaged pelts that led, in part, to the widespread use of the gin trap (the name 'gin' is derived from the word 'engines' as the traps used to be known). These were designed to catch and hold by the legs, not to kill, thus preventing damage to the skin. Though undoubtedly effective, they were cruel in the extreme, and were quite rightly banned in 1958. The search was now on for a replacement humane rabbit trap, one that would kill quickly, and could be set within the rabbit burrow, thus minimising the chance of catching non-target species. Many designs were tried before they settled for the Imbra and

The Imbra trap (left) and the Juby trap

Transport your Mk 6 Fenn traps with the safety catch engaged

Juby type traps. These work with a scissor-like action. I've found them very efficient and they catch plenty of rabbits, but the drawback is that they are quite heavy and cumbersome if you have to transport them any distance. I only use Mk6 Fenn traps nowadays; they are easy to set and light to carry if you have to transport them far. So, how do we go about setting Mk6 Fenn traps for rabbit control?

Very occasionally as a pest controller, you may be asked to clear a large warren of rabbits on a regular basis. When you have removed the residents with an all-out assault by the ferreting team and filled in all the holes, many rabbits that may have been sitting out while the operation was in progress, will, in a short while, open up certain holes and start to re-colonise the warren. These reopened or 'active' holes are wonderful places for catching rabbits. This is the perfect situation where the careful setting of Fenn traps can, at times, bring some spectacular results. Before setting the traps, you must clear all the vegetation from around the holes you intend to trap because this makes the job a lot easier when the time comes for inspection and resetting. We'll assume the traps have been sufficiently weathered and are in a serviceable condition; by that I mean the spring mechanism is operating with a strong and smooth action. If necessary, use a little light lubrication such as clear candle wax or vegetable oil to rejuvenate a corroded trap. There are a few other items you will need before setting the traps. They are a trapping hammer, a small trowel, and a bucket of sieved earth for covering the traps after setting. You will also need a setting stick, which is a length of straight hazel about 12 mm thick with a tapered shape at one end, to slide under the trigger plate to hold it steady when covering the trap with earth. When transporting a lot of Fenn traps, it pays to set them up with the safety catch engaged before leaving home, this way you can carry a larger amount and you haven't got to try to set them up in the field with freezing muddy fingers (I like to carry my traps threaded onto a loop of stout cord which I carry over my shoulder). Right then, we are about ready to start.

Select the first hole and, with the trowel or trapping hammer, scrape out a bed for the Fenn about 8 inches inside the hole. It should be deep enough to allow the trigger plate to sit level with the floor when set and also allow just enough room for the jaws to snap shut without touching the roof of the hole. Place the unset Fenn into the bed you've just scraped out and check it sits snugly without rocking about. When you are satisfied it fits comfortably, remove the Fenn and engage the trigger plate, and set it as fine as you can, **keeping the safety catch on.** Using the trapping hammer, knock in the holding peg to one side of the hole, making sure it goes in well out of sight. Place the Fenn back into the bed, keeping the jaws of the trap to the sides of the hole, and carefully slide the setting stick under the trigger plate. You can now gently cover the trap with a couple of handfuls of the sieved earth from your bucket, keeping it as level as you can. Gently withdraw the setting stick and smooth over the hole the stick came from. Now, **using the setting stick**, very carefully move the safety catch to leave the trap in a set position. The trap is now ready to catch. The reason you use sieved earth to cover the trap is to make sure there are no stones in the soil to impede the working of the trap. One other important point is to leave the soil covering the trap level or slightly higher than the floor of the hole; if you leave a depression, the rabbit will jump over it and miss the trap. When you have set your first trap, stand back and look into the hole. No part of the trap should be visible and it must appear perfectly natural. Once you are satisfied with the setting you can move on to the next hole. You then repeat the procedure for the remaining holes in the warren, which can range from four or five holes to forty or more, depending on the volume of rabbits. All that's left to do is to check the traps often, at least twice daily; first light and dusk are as good a time as any. Hopefully your labours will have been rewarded with a heavy catch; a lot of young rabbits can be caught in the summer months using spring traps which dramatically reduce the breeding potential later in the year. If a trap doesn't catch for a day or two, carefully remove it from the hole and set it somewhere else, making sure you block the hole you took it from. I can't stress enough the importance of using the safety catch when setting traps. Twice I have had a Mk6 Fenn go off on my fingers and, believe me, it's an extremely painful way of remembering to engage the safety catch. That said, don't be afraid of the traps. They are very safe and easy to use if all the safety aspects are followed carefully, and the more you handle them the more confident you will become. If you intend doing a lot of trapping it may be worthwhile investing in a pair of knee pads because you will be spending a lot of time kneeling down (try setting forty Fenns without pads and you will see what I mean).

In today's hi-tech world it's easy to forget just how effective the old-fashioned trapping techniques can be. It's a skill that once mastered can be used in countless other situations, and although you may not use it often, it's a handy weapon to have in ones armoury. That's basically all there is to spring trapping, simple but very effective in the right hands. As always, experiment a little with different settings and locations. Sometimes the odd bolt hole that comes from an inaccessible bury on the other side of a rabbit proof fence will produce rabbit after rabbit, day after day, so always be on the lookout for new trapping opportunities.

Gassing

Since the demise of that faithful old servant Cymag, the only gassing compound available for use nowadays is aluminium phosphide. It goes under the trade names of either Talunex or Phostoxin and is applied, in both cases, by means of a Perspex tube which dispenses pellets into the holes via a trigger mechanism. I won't go into the details of application for this simple reason; anybody wishing to buy and use aluminium phosphide **must** have been on a recognised training course to learn how to use this potentially lethal substance safely and efficiently. The BPCA (British Pest Control Association) runs training courses that cover the use of aluminium phosphide, and I strongly recommend anyone thinking of taking up full time rabbit control to go on one. Gassing rabbits should never be carried out by one person alone; it always requires two or more people to administer the substance safely. Gassing rabbits is not the miracle cure for rabbit infestations that most farmers and landowners think it is. So many factors need taking into consideration before deciding if gassing is the best option. Are the buries shallow or not very deep? Is the soil sandy and porous? Is there a watercourse adjacent to the buries? Many of these things would rule out the use of gas as an option, so go on a training course to learn where, and how, to use it safely for the best results.

4
Mole

I get as many calls to control moles as I do rabbits in the winter months. For some strange reason Mr Mole, or rather the result of his activities seems to bring out the worst in certain people. Normally level-headed customers are sometimes reduced to dangerous psychopathic demons after fruitless attempts to rid their beloved lawns and gardens of this loveable little rogue. If there is one animal guaranteed to make the pest controller look utterly foolish then the common mole (*Talpa europea*) is definitely top of the list. It can be simplicity itself at times to trap him but often he proves to be an elusive character to even the most experienced of mole catchers, myself included.

Let's start at the very beginning and look closely at the mysterious mole and his subterranean lifestyle. The molehills that greet us when we arrive at a client's house are the result of his constant tunnelling activities in pursuit of his main food source, the humble earthworm. These tunnels are merely worm traps and they work on the principal that a worm travelling through the soil will, sooner or later, find itself emerging into one of the mole runs. The worm then proceeds to crawl along the run until such time as a foraging mole finds it. The mole works as much in the daytime as it does at night, which comes as a complete surprise to a lot of people. It is supposed to work at approximately four hourly intervals, alternating between rest and work (a bit like some people I know). The truth is, like most animals, they hunt when they are hungry and sleep when they are tired. Although the moles eyesight is decidedly poor, it is not blind as many people suppose, it makes up for the lack of sight with an acute sense of hearing. There are no external ears yet his hearing is very sensitive, with the soil assisting the mole in picking up even the slightest vibration. Try walking up to a mole that is working and see how quickly it disappears. There is always debate as to whether the mole has a good sense of smell; some say he has, others disagree. I am of the opinion that his sense of smell is the same as any other small mammal, and for this reason I will continue to weather my traps before use.

Types of Trap

The main types of traps in use today are the pincer or tongs type and the Duffus or half barrel trap. There are live catch traps available but, unless you are on hand to release the trapped mole straight away they can be less humane than the instant kill type. Another trap that is available is the Talpex. It is a pincer type trap that has a very powerful spring mechanism, and I have started using these just lately to ensure a quick and humane kill. Whichever trap you decide to use it must be powerful enough to do the job. At the risk of repeating myself, all traps need to be weathered before you attempt to use them and mole traps want burying in the compost heap for a few weeks prior to use. It always amazes me to see people use brand new traps on a job without preparing them first, and then wonder why they never catch anything.

The art of trapping any animal always comes down to the skill of the trapper. Knowing where to set a trap is as important as knowing how to set it. With mole trapping it is vital to set the traps on an established run in an area of fresh activity. I always look for a run about four or five inches deep, probing between two fresh molehills using a dibber bar. Sometimes it's not possible to locate the deeper runs and you are forced to set in the shallower ones, but as a general rule, the deeper the run, the more established it is. When you have found your run you need to expose it (I use a bricklayers pointing trowel), making sure you open up the hole leaving plenty of room for the trap to spring.

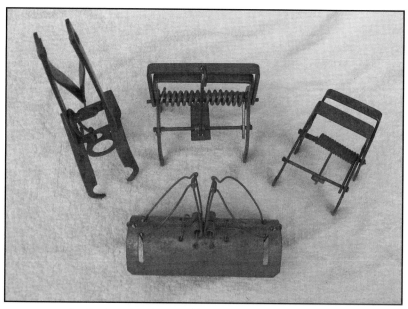

A selection of mole traps (the Talpex trap is centre back)

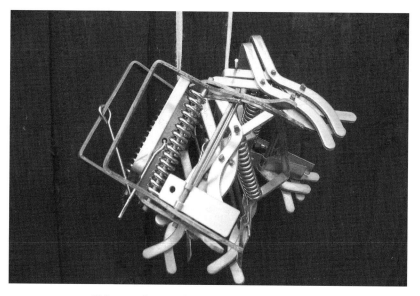

Talpex mole traps hanging outside to 'weather'

Remove a square of turf wider than the sprung trap if using the tongs type, because this way you can be sure the trap will spring effectively. With the tunnel exposed, remove as much loose earth as you can, so the run remains as smooth and undisturbed as possible. I use a short length of an old T-shaped spade handle with one half of the T sawn off; I use this for tamping down any loose soil in the run. Before inserting the trap in the run, I like to rub some loose soil taken from the hole all over the trap to help mask any smells; indeed, I always rub my hands in the nearest fresh molehill prior to starting trapping for the same reason. With the tongs-type trap, I usually fill the ring of the trigger mechanism with moist earth so when the mole contacts the trap, he encounters earth, not metal which would arouse suspicion. With the trigger set as fine as possible, you can position the trap within the tunnel, rocking it from side to side to ensure a snug fit and to make sure there are no stones or roots that may impede the closing action. That's it; all you have left to do is split the square of turf in half (the one you removed when you started), place half each side of the trap, and cover with a light sprinkling of earth to exclude any light. The trap is now ready to catch. One last thing, push a marker stick into the ground next to the trap to aid location later on. Many traps are lost each year because the trapper had forgotten where they'd been set. If you are trapping on a golf course or playing field it is essential that you mark the trap locations. These areas are mown frequently and a trap being hit by a set of gang mowers can do an awful lot of damage. I mark my traps with yellow sticks that have red and white bands around the top which can be seen easily and avoided by the mowers.

When attending any new mole job, be aware that the client has probably been trying to get rid of the little blighters for ages, prior to calling you in. Worse still, they may have had another contractor in before you, trying to catch them as well. Every remedy known to man would have been tried before you get there; mothballs, creosote, and Jeyes fluid shoved down the holes is common, as are windmills, bottles and sonic devices stuck in the ground. You name it and I've seen it. All this does is make the mole ten times harder for you to catch. Always point out to a prospective client that when others have tried to catch a mole before you arrive, it is often very difficult to achieve a successful outcome. If a trap has been triggered and the mole missed, it can be extremely difficult to tempt a mole back through a trap again, preferring instead, to earth it up. Finding traps earthed up are a frequent occurrence anyway, and when this happens, the best solution is to change the type of trap you are using or to move the trap somewhere else. Other things to look out for when setting traps on unknown ground are sharp objects under the turf like tin or glass. I have an inch long scar on the back of one of my fingers from a nasty cut that required several stitches. This was the result of a broken bottle in the ground, so be very careful.

When mole trapping don't expect to catch a mole at every visit. Like all trapping, you can have good days and bad; sometimes the mole moves away to work another area before returning to where your trap is set, so it may take several days for a trap to spring. The fastest I have ever caught a mole after setting a trap was about two minutes. I had just finished setting the first trap in an orchard and was probing around with the dibber bar looking for the next run, when I noticed the first trap had sprung. Thinking I had set the trigger too fine, I pulled it from the ground to reset it, only to find a mole caught in the jaws. As I knew there was only one mole in the orchard I took up the trap, presented an astonished client with the bill (and the mole), and set off for home. Now this was a complete fluke. Normally it takes at least a couple of visits to remove a single mole, and often more than that, but if there are a lot of moles, then you can sometimes be going back for weeks. If the area you are trapping is quite a distance away from home, ask the client if they would mind looking at the traps for you. They can give you a ring if one has gone off, saving them and you a bit of money in the process.

We have discussed the setting of the pincer type traps; the one other type we need to look at is the Duffus or half barrel trap. These I find useful if setting in an area where discretion is needed, like a school

Damaged mole trap, the result of being hit by a lawnmower

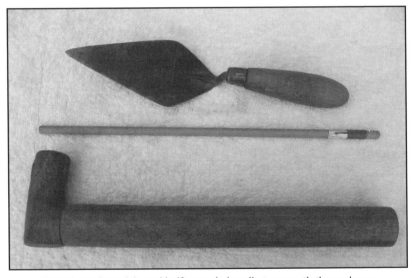

Trowel, marker stick, and half a spade handle to smooth the mole run playing field for example, where traps would soon go missing if they were projecting out of the ground. Also, you don't want prosecuting because little Johnny has tripped over one of your traps and broken his leg! Duffus traps when new require a bit of fine tuning to make them work more efficiently. Straighten the trigger catch that sits on top of the trap a little with a pair of long nosed pliers and the trap will then be able to fire on a hair-pin trigger. Bend the trigger wires that hang underneath the trap slightly outwards so they don't touch each other, and also widen the gap between the wires a little bit. They will now be ready for setting. The turf needs cutting out the same length as the trap, and the run clearing out the same as for the setting of a pincer trap. The Duffus trap is then very gently bedded down into the run and carefully covered with loose soil (not turf) from a nearby molehill. You can leave the loose turf next to the trap as a marker for its location. If you push it down into the ground a little with your foot, it will pass virtually unnoticed to the untrained eye. The beauty of the Duffus trap is that it can catch two moles at a time; I have done it many times myself in the past, although it is not a very common occurrence. Everyone will develop their own method of setting mole traps; no two trappers will set in exactly the same way, so find a method that works for you and seek to perfect it. Experiment with different settings and locations, and if one type of trap is not catching, switching to another can often pay dividends. Occasionally you will find oddities in your mole traps. Weasels are often caught as they hunt the mole runs, and I am always saddened a little when I catch one of these because they were helping me with the job so to speak. Sometimes you will catch moles of a different colour. I once caught three apricot coloured moles from the same garden in the space of a month; they were obviously all from the same family. I had one of them sent to a taxidermist so I could have a permanent reminder of the remarkable achievement. Funnily enough, I have been back to the garden many times since and have only ever caught normal coloured moles. White moles occur from time to time. I have seen a stuffed specimen but I have never been lucky enough to catch one myself. I have also seen pictures of part coloured moles, and one was a normal colour on top with a fiery red underside. On the

One of three apricot coloured moles caught from the same garden

Always mark the location of your mole traps so you don't lose them

colour of moles they are actually dark grey in colour and not black as many people suppose. Don't believe me? Next time you catch one brush his coat aside and have a look.

Love him or loathe him the mole is a fascinating and mysterious little creature. He has a long and colourful history and has long been the subject of myth and legends. A common superstition was that to carry the dried front feet of a mole in one's pocket would ward off rheumatism (a custom still carried on in parts of East Anglia). The mole was also responsible for a famous royal catastrophe, causing the accidental death of William III, Prince of Orange. In 1702 the King's horse (named Sorrell), stumbled on a molehill whilst being ridden in Hampton Court Park, and the King, who fell from his horse, died of his injuries sixteen days later. The molehill is portrayed at the base of the statue of the King in St James's Square in London. As a result of his exploits, the mole was held in high esteem by the Jacobites and a favourite toast of theirs was to 'The Little Gentleman in Black Velvet'.

Since the last outbreak of foot and mouth disease, when rural pest controllers were unable to gain access to vast areas of countryside, there has been a massive rise in the mole population. Combined with the exceptionally wet and mild winters of late, this has led to the almost unprecedented mole numbers we see today. So if ever there was a good time to learn the art of mole trapping, then this must be it. Don't give up if you are not successful for the first few attempts with the traps, practice makes perfect as they say, and if you persevere you will become proficient in the end.

Gassing

Moles can be gassed using aluminium phosphide tablets the same as in rabbit control.

But as in rabbit control, the user must have been on a recognised training course before he can purchase or apply this potentially dangerous substance. It can be effective in certain situations; indeed I use it myself at times, but usually only as a last resort if all other methods have failed.

Strychnine Hydrochloride

At the time of writing, strychnine is being withdrawn from the market due to a European directive. The BPCA (British Pest Control Association) and the NPTA (National Pest Technicians Association) have both submitted applications to have the ruling overturned but as yet no firm decision has been made. As a control method on large acreages it was undoubtedly the most cost effective and successful method of mole eradication available.

5
Grey Squirrel

With many people now calling for a mass cull on the grey squirrel numbers it may be a good time to examine the many and varied methods of squirrel control in the rural environment. Like most pest controllers, the majority of calls I get regarding the grey squirrel is to remove them from roof spaces. They can cause immense structural damage if left to their own devices. I have seen ceiling and roof joists gnawed through to the point of near total collapse and, more dangerous still, electrical wiring stripped bare of its insulation so that it posed a major fire risk. In an urban situation there is a never ending supply of squirrels, and like foxes, if you remove one, another quickly takes over the vacant territory. In nature, a vacuum isn't tolerated for very long. I have seen proof of this on many occasions and it applies to magpies, crows, foxes, pigeons, etc. You remove a few and in a short while the area is re-colonised. Total eradication of the grey squirrel is now virtually impossible; like the rabbit he is here to stay. All we can do is hope to keep on top of the situation. Like a lot of animals originally from a rural environment, the squirrel has quickly adapted to an urban existence. A freshly replenished food supply from bird tables and bins means he has no need to struggle to find food in his natural habitat. I have a few contracts that require me to remove the greys from woodland areas, and in these I have waged all-out war on them for some time with varying degrees of success. Now I must stress, the areas I control contain **no** red squirrels, so I can use all methods of control on the greys. If the places you intend to work contain red squirrels, then only live catch traps can be used for obvious reasons. An adult grey squirrel weighs about 450 g and its muscular body, excellent sight and special eye structure adapt it well for its mainly arboreal existence. The squirrel though, is equally at home on the woodland floor, in gardens and urban parks. Squirrels are basically diurnal creatures, with the main peaks of activity at dawn and just before dusk. In bad weather squirrels may remain in the dreys for long periods as they are particularly susceptible to fatal chilling when wet. Therefore, as pest controllers, a good time to attend to the dreys would be in periods of cold wet weather. Early spring (March/April) is a good time to sort out the dreys, just before the period of maximum tree damage. Let's take a look at all the options open to us.

Spring Traps

One of the best spring traps of the instant kill type for grey squirrels is the Kania. They are a little costly to buy initially but are extremely efficient in effecting a clean kill every time. I use a few Kanias permanently set high up on the trunks of the squirrel's favourite trees. Baited with peanut butter they prove irresistible to the ever inquisitive squirrel. They enter from underneath the trap and the very powerful spring mechanism ensures a quick and humane end to every visitor. I camouflage my traps with a little vegetation to help them blend into the landscape and to prevent unwanted attention from well meaning (sometimes) members of the public. If trapping in areas that have public access it is always beneficial to disguise traps and cages. Not every person shares your attitude to vermin control, and you can often find your expensive traps have been damaged or worse still, stolen.

Wooden tunnels containing Mk4 Fenn traps placed at the base of well-used trees are another highly successful method of clearing large numbers of squirrels. Baited with cobs of maize they will account for many squirrels especially in the autumn months. During the autumn, squirrels are very active in searching for food to store away to see them through the cold winter months. A tunnel baited with maize will prove very effective in luring the squirrel over the trigger plate of the trap. I

The Kania trap will catch regularly if fixed to a favourite tree

I prefer the .177 air rifle over the shotgun for woodland squirrel control

use a tunnel with only one entrance open, the Fenn trap is placed in the centre of the tunnel bedded slightly into the ground, and the bait placed at the blocked end. The trap doesn't have to be covered but you can if you wish place a few leaves over it to disguise it a little. The entrance holes to my tunnels are only large enough to allow access to smaller mammals such as rats, stoats, squirrels, etc. This prevents the possibility of larger animals or birds (pheasants for example) gaining access to the bait. Pre-baiting an area prior to setting the tunnel traps can greatly increase your chances of success. It gets all the squirrels in the immediate vicinity used to feeding in exactly the place you want them, and if you leave a few tunnels without traps dotted around, they will soon accept them as part of the scenery. Squirrels simply cannot resist inspecting new objects, especially ones with nice dark holes. For this reason setting tunnel traps for them is one of the most productive methods of control.

Drey Poking

One of the more exiting methods of squirrel control is the poking out of the dreys in the early spring, before too much cover on the trees makes them difficult to locate. It requires two, or better still three people, to carry it out successfully. One person carries a set of lofting poles (the sort used in pigeon shooting) to 'poke' the dreys, the others stand around the trees with shotguns waiting to shoot the bolting squirrels. Great sport can be had at times; a bolting squirrel can prove an elusive target as it always tries to position itself on the side of the tree furthest from the guns, hence the reason for having three guns instead of two. Sometimes the dreys are too high up in the tree canopy to reach with the lofting poles so a blast from the choke barrel is needed to get the squirrels moving. If you can sort the dreys out in periods of cold wet weather, you are more likely to find them occupied. March or April is a good time to attend to the dreys. You will be disrupting the main breeding period at this time and so dramatically restrict the potential for the squirrels to reproduce in numbers. I regularly walk around the woods every couple of weeks or so from February to May with the shotgun and systematically blast every drey I see. It keeps the squirrels constantly moving and eventually leads to the dreys being completely abandoned. Some squirrel dreys are quite substantial constructions and it may take a lot of poking with the poles or blasting with the shotgun to render them uninhabitable. You will be amazed at some of the things the squirrels use to construct their tree-top nesting places. Apart from the stripped bark fibres, you can find polythene bags, sweet wrappers, papers, tin foil, and on one memorable occasion I found the remnants of a pair of ladies tights. I can't begin to imagine from where they were acquired.

Anticoagulant Bait

Another method of squirrel control that used to be quite popular was the dispensing of warfarin-laced grain from specially designed galvanised steel hoppers that only allowed access to the grey squirrel. I was never a fan of this method as I always tried to keep the use of anticoagulants to a bare minimum. I would always worry that other animals may feed on the squirrel carcases, so for that reason I never used this method. In remote areas of woodland with limited public access, where the frequent checking of traps is not possible, the use of poison-bait hoppers does have its merits, but this would be the only place I would consider using it.

Cage Trapping

Cage traps, be they multi-catch or single-catch types, are still one of the best control methods for a secluded woodland situation. Baited with peanuts or maize they are highly visible from high up in the tree canopy and squirrels can seldom refuse the chance of a free meal. Picnic areas in woodland that see lots of activity during the summer months can prove excellent trapping places in the winter, because squirrels have

Squirrels can be caught in cages where they regularly come to feed

Squirrels will do immense damage to get at a favourite food source

been used to feeding there on a regular basis. The only problem with cages is, that if they are used where the public have access then they frequently get damaged or stolen, which being expensive to replace, is not something you want happening too often. You can disguise them with a little vegetation, but, if the public find it hard to spot them, so will the squirrels, which is not what you want. That said, cages are excellent for catching large numbers of squirrels; they seem to enter them with very little hesitation and, the beauty is, any non-target species can be released unharmed. If you are setting cages on the woodland floor, then pin them down with a T bar or stick pushed through the wire to prevent foxes or badgers walking off with the cages. For a novel use of a cage trap, I had a client who wanted me to catch some squirrels that had eaten through the lid of his industrial-sized waste bin to get at the food scraps inside. I placed a cage vertically directly beneath the hole and a mouse bait tray full of peanuts at the bottom. It caught six squirrels in as many days, much to the customer's delight. Multi-catch cages are sometimes used in large areas of woodland; these can be very successful at times as the trapped squirrels encourage the more hesitant into the trap. Baited with maize or peanuts, they entice the ever inquisitive squirrels through the double non-return doors into the large holding chamber. The most squirrels I ever had at one time in a multi-catch cage was seven, two females and five youngsters. I have heard of larger catches than that in heavily populated areas. Again, pre-baiting an area for a week or so prior to installing the cages can, in most cases, dramatically increase your chances of success. The multi-catch cages can be used in roof spaces though I have only had limited success when using them there. Single-catch cages are much more productive. Whilst on the subject of squirrels in roof spaces, make sure you place the cage on a flat board or something similar, and that the cage is weighted down with a brick. Also, you should not place it near electrical wiring or stored items because a trapped squirrel will pull anything it can get hold of into the cage with possibly disastrous consequences.

There is often a problem when a squirrel has a nest of youngsters in an inaccessible place, behind the fascia board for instance. If they have access to the roof space and are weaned and well developed, they will readily enter the cage especially if the mother has been caught first (use the same cage in this instance). The difficulty arises when the young are too small to leave the nest. The female is usually easy to catch, but you may have to remove a few roof tiles to remove the young. A word of warning here though. If you attempt to remove the litter and the female is still about, she will be alerted by their alarm calls and come running to defend her young. I have never been physically attacked by an irate squirrel but have come close on many occasions. I was once treating a property for ants when the mobile phone rang, asking me to come and remove a nest of squirrels from inside a blocked-off chimney to a bungalow. The gentleman was having a new TV aerial fitted and the chaps doing the installation would not go back on the roof because the female squirrel, alerted by the alarm calls of her young, would not let them near the chimney. I duly arrived at the house to find two fitters locked in their van and a very agitated customer standing on the drive gazing skyward. "Got a roof ladder?" I enquired, "Na, we walk up the tiles to the ridge" replied one of the fitters, having been persuaded to leave the sanctuary of his vehicle. Great I thought, so sticking the air pistol in my waistband and grabbing a torch, I gingerly made my way up the tiles to the ridge. Shining my torch down the chimney pot I could count three well-developed youngsters about four feet down in a nest. I loaded the air pistol and dispatched the first squirrel; the remaining two began their distress calls and, along the garden fence and up the roof came the mother. I took a pace back and was now delicately balancing on the ridge unable to reach the pistol because the squirrel was sitting on it. Every time I inched forward she threatened me with her chirring warning call. To anybody watching from a distance it must have appeared as some bizarre dance ritual but if I didn't get a hand on the chimney pot soon there was a fair chance I was going to lose my balance. It was now a 'toss up' as to who was going to get killed first, me or the squirrel. After what seemed an age, she eventually disappeared down the chimney pot to check on her litter and I was able to grasp the chimney at long last and regain my balance. A second shot down the chimney missed its target but the noise startled the female and she abandoned the chimney and sought refuge in a nearby tree, still chirring defiance. The remaining squirrels were quickly dispatched and removed from the chimney pot and a cowl fitted to prevent any further squirrel problems. The fitters, who had watched the whole procedure from the safety of their vehicle, could now continue fitting the TV aerial, much to the relief of the client. This shows what lengths some animals will go to protect their offspring. It doesn't only apply to squirrels; many other animals and birds also will not hesitate to attack in defence of their young, so always proceed with extreme caution when dealing with this type of situation.

Shooting

Apart from the shooting that ensues from drey poking, it can be productive to walk through woodland in the winter and spring months with a shotgun or better still an air rifle, and remove the foraging squirrels. In autumn the squirrels are very active collecting food to see them through the coming winter. They spend a lot of time on the ground during this period and can make easy targets when they stop and bury food. Also, the tree canopies have yet to fully emerge and the squirrels will be quite visible in the tree tops, again making an inviting target. I prefer the air rifle over a shotgun since it doesn't disturb the squirrels as much. In areas that are frequently shot, squirrels will often seek refuge in a convenient hole in a tree until the danger has passed. I don't recommend the use of the .22 rimfire because of its potential to ricochet and also you don't know where or how far the bullet will travel if you miss the target. Most modern air rifles are quite capable of dispatching squirrels at distances of 35 metres and more, so this would be my preferred option. Whilst mentioning dispatching squirrels, the air rifle is the best weapon for dealing with small cage trapped animals, including mink. A shot to the head from close range and death is instantaneous. You can use the .22 rimfire if you wish, but the cage must be on soft earth and you need to use the lower powered CB caps. Often when shooting tree-borne squirrels it's handy to have a couple of friends with you so they can cover all sides of the trees. As previously stated, the squirrel will always try to position itself on the side of the trunk furthest from danger. The squirrel is a tough little animal (a lot tougher than they appear), so an accurate shot is required to bring them to book. Head shots are nearly always successful but an area targeted just behind the front elbow will prove equally effective in making a clean kill. I like the penetrative power of the .177 over the .22 air rifle when stalking squirrels, as a carefully placed shot should **always** ensure a quick and humane end. If you cannot guarantee to hit the target, then don't fire. Be aware of the sensitivities of other people if shooting squirrels where you can be observed by the public. The squirrel you just shot might be the one that a few moments earlier was being fed by a little old lady in her garden. Discretion is paramount when dealing with squirrels so you should always pick up any bodies and take them away for disposal later. I use a large game bag for this very purpose. Often white lies are needed when confronted with concerned members of the public. I sometimes say that rats have been reported and I have been called in to deal with them, which is not really a lie. Squirrels are tree rats, aren't they?

6
Mink

The mink is rightly regarded as one of the most prolific predators at large in the countryside today; garden ponds, fisheries, farms, and poultry pens are frequently invaded by this ferocious killer who will often attack prey much larger than itself. As a result of deliberate releases and escapees from fur farms, we now have mink breeding quite freely along waterways and around lakes throughout most of Britain. If I got a call from a customer worried that fish were disappearing from a pond that was netted over, I could be fairly confident that a mink was responsible. If a closer inspection revealed half eaten fish or frogs stored under overhanging vegetation, then I would be certain that a mink was active in the area. The mink is a secretive animal and is seldom seen, but they hunt as much in the daytime as they do at night. It's because they are rarely witnessed hunting their prey that some people believe they are difficult to trap. Of all the pest species you will be asked to trap, the mink is probably the easiest to catch; being quite a lazy creature it rarely refuses the chance of a free meal and, for this reason, are readily caught in cage traps. Mink feed on fish, frogs, birds, and small mammals, with rabbit being the most popular. They are equally at home on land or water, but most of their time is spent hunting the water margins. It's here that a carefully placed cage trap will produce the best results.

Cage Trapping

Mink cage traps need to be robustly constructed. A 12-gauge mesh is required as a minimum because a trapped mink will test the strength of

73

any cage to the limit. I have seen lesser gauge cages ripped apart by an irate captive mink. As with all cage trapping the scent of a previous victim can prove irresistible to the next visitor to the cage, so for this reason I never clean my cages once they have caught. New cages have to be weathered before they are ready to use for trapping, as we've said previously and it is even more important in mink control as they, like rats, are sometimes neophobic (scared of something new). The most popular bait for mink has to be rabbit, with pigeon breast (complete with a few white feathers) coming a close second. In reality, any smelly or bloody bait will suffice; I have had good results with pilchards, kippers, liver and cat food in the past. As with all things, experiment a little with baits because certain baits perform better in different areas. Find the bait that gives you the best results and stick with it. Mink have a very acute sense of smell and will scent a baited cage from an amazing distance away and as pest controllers we can use this to our advantage. I was shown a trick by a gamekeeper acquaintance of mine who used to drag an old piece of sacking or rag, soaked in a mixture of rabbit blood and fish oil (the oil from a tin of sardines will do) around the perimeter of his ponds and up to the entrance of his tunnel traps. He had some very good results using this method.

The positioning of cage traps is very important since they must be placed where a mink will discover them easily. If you can find where a mink runs along a water margin then this is one of the best places to set a mink trap. If you can find an area of level ground under some overhanging vegetation so much the better. If needs be, use a trowel or spade to level an area of ground, then bed the cage down into it, not forgetting to push a metal rod through the wire of the cage to hold it tight to the floor. This also prevents the cage from toppling into deep water should it become dislodged by a trapped mink. If you can cover the cage with vegetation, it will not only conceal it from members of the public, but make it far more inviting to a foraging mink; they cannot resist exploring dark inviting holes. When baiting the cage, wire the bait into the trap in such a way that the mink cannot reach it from the outside. If it is close to the edge, the mink will try and pull it through the wire without entering the cage. As with rabbit trapping, you can position more than one cage along the same run but have the entrances pointing in opposite directions. That way, from whichever direction the mink comes along the run, he will stumble upon the inviting entrance to a cage. This is particularly effective if you have a bitch mink with young. Once you have caught the mother, the youngsters will quickly follow her scent into the cages. As with all trapping the cages must be inspected at least twice a day, and any occupants humanely dispatched. On the subject of humane dispatch, you should never attempt to drown a mink in a cage, it is both cruel and inhumane, and we owe it to whatever species we trap to see that its demise is as quick and stress free as possible. All mink should be shot at close range with a suitably powered air rifle or .22 rimfire using the lower power CB cartridge. Death should be instantaneous. On a lighter note, take care when handling mink in a cage, as a carelessly placed finger can result in a savage bite. I have never been bitten by a mink, but I have received a severe bite from an irate hob ferret; it's not something I would care to repeat in a hurry.

If you are called to deal with a mink that has killed poultry and has left some dead birds behind, it will invariably return within a night or so to retrieve the remaining birds. The same is true if a stoat or a fox is the culprit as they nearly always return to the scene of the crime. A cage placed just inside the entry point baited with one of the dead birds will often account for the villain responsible. On more than one occasion I have trapped two or more mink from the same location although these are usually immature youngsters from the same litter that have yet to claim their own territory. When dealing with a suspected mink around a garden pond look for the nearest harbourage. This could be underneath the shed, in the filtration system to the pond, or anything else close by that could conceal the unwanted visitor. If you can find a likely hiding place (and it will be very near the pond), place one cage near the cover and one alongside the pond. Always try to maximise your chances of success. It pays to weight the cage down with a heavy log or stone to prevent the powerful mink from overturning the cage in an attempt to get at the bait, and disguise it a little with some vegetation so the trap becomes a dark and inviting place for the mink to investigate. Try and set the cages in a secluded area around the pond, somewhere with little disturbance; mink like to feed where they think they can't be seen. I usually find that a client has witnessed the mink travelling to or from his favourite feeding places at least once or twice, so acting on this information it should give you a good idea where the cage wants placing for the best results. You may also find where the mink comes in and out of the water or where it has been sitting to devour its catch so placing a trap here will almost guarantee you successful results. As with rabbit trapping or squirrel trapping, it is quite normal that the cages may not catch for a few days or a least until the minks' suspicions are allayed. That said, I have caught a mink within hours of setting a cage trap, so you never can tell. One other thing to remember when trapping near water is that rats frequent these areas as well, so be aware of handling contaminated vegetation or soil as rat urine can contain the bacterium that causes leptospirosis (or as it is better known, Weil's disease). It is sensible to wear protective gloves when working anywhere near a rat infested water course for this reason. Weil's disease will be discussed further in the rodent section.

Spring Traps

Apart from cage traps, mink can also be caught in spring traps. The Fenn Mk6 is the most popular (the Mk4 is not powerful enough to ensure an instant kill); these can be sited in tunnels or holes in the bank at the waters edge. The tunnel trap can be built as a permanent structure along a well-used mink run and if it is sturdily constructed from stone or logs it will give many years service before having to be replaced. Just above the water line is where mink (and rats) like to travel and a trap placed here will intercept both species as they forage in search of food. Suitably disguised, they soon blend into the surrounding vegetation and will pass virtually unnoticed to all but the trap builder himself. If you can find (or make) a narrow bridge over a ditch or brook, then these can be deadly trapping places. A narrow log or length of wood is ideal. If using a log, an area needs to be chopped out with an axe to leave a flat section for the trap to sit on. The chain of the trap wants fixing underneath the log with a staple to become a permanent fixture. The trap must be covered with a tunnel made from wire mesh bent to the required shape so as to prevent the capture of non-target species. If the wire tunnel is then draped with quantities of heavy vegetation it will be held securely in place until such time as you need to inspect it. The wire tunnel wants to be tall enough to allow the trap to spring shut leaving about half an inch clearance at the top and it wants to be wide enough to accept the set trap with the safety catch disengaged. The tunnel needs to be left in place for a few weeks without a trap in position, to allow the mink to use the bridge with confidence. If you have sited the bridge in the correct place, it should soon become a favoured crossing place not only for the mink but many other animals as well. I have caught rats, stoats, squirrels and even rabbits in the ones I have built, so they really do work. The correct location is vital though as it may take several moves of the bridge before you find the most productive place. Having said that, moving the bridge every now and then can often increase you chances of success anyway. The places to site these bridges is where you can see signs of something regularly jumping the water (usually at a narrow point in the ditch or brook), but you can force animals to use your tunnel by the careful placing of wire netting put across and blocking their normal runs.

If you have a recurring mink problem, around a lake or fishery for example, then Mk6 Fenn's placed in specially constructed holes cut into the bank at water level, can prove very effective in catching a newly arrived visitor. Any wandering mink new to an area, will seek to investigate every nook and cranny in its bid to set up a fresh territory. If you have ditches or streams that run into the lakes then these will be the access points the mink will use to reach the fisheries. If traps are placed both sides of every ditch or brook that runs into the lakes, then any foraging mink will be intercepted before it can do much damage. The holes want digging into the banks at water level; they need to slope gently upwards for a depth of about 18 inches to a shelf that sits out of the water right at the back of the hole. The water wants to lie in the hole to a depth that just covers the trigger plate of a set Fenn trap (about 1 inch). The bait needs placing on the shelf at the back of the hole (a rabbit liver is as good as anything), before the Mk6 Fenn is placed in the hole just under the overhang of the roof. The roof of the hole must be high enough to allow the jaws of the trap to spring shut. When the trap is gently bedded into the base of the hole the pan should be sitting just under the water level. Finally, a stick about 12 mm thick needs to be placed across the entrance to the hole about 1 inch above the water level, just in front of the set trap. The idea is that a mink will smell the bait at the rear of the hole, jump over the stick to investigate and land with its full weight on the trigger plate of the trap. The ends of the stick want pushing into the sides of the hole to help keep it in place. If these holes are positioned on opposite sides of a ditch or brook about twenty feet before they enter the lake, this is as good a position as any to achieve maximum results. If there are large diameter pipes running into the lakes then these are also good locations for traps. Dig the holes either side of the pipe in this case. Another excellent place for catching mink is in a rabbit bury if one happens to be on or near the bank of a lake. I discovered this quite by chance when trapping rabbits on the banks of a private fishery. I put ten Mk6 Fenns in a fourteen-hole bury and in the space of a fortnight had caught nineteen rabbits and four mink, much to the surprise of the owner who had been unaware he had a mink problem until then. Other good locations to trap mink are around the perimeter fences of pheasant release pens. Mk6 Fenns placed in tunnels will account for many rats and stoats as well as mink. If you are working in an area with signs of water vole activity then you **must not** use spring traps for controlling mink.

Drainage pipes are good locations for mink cage traps

Natural or artificial bridges are ideal places for cage or tunnel traps

Shooting

Mink can be shot quite readily where they appear in any numbers; you can either sit out and wait for them to appear, if they are known to be in a certain area, or you can quietly stalk the waterside and hope to catch one unawares. The twelve bore shotgun is the preferred weapon here, as often a mink will try to escape by swimming across open water, especially if it is being pursued by a dog. They offer an easy target when in the water. Not so when on the banks though, as they can be extremely difficult to hit as they weave their way through the bank side vegetation at high speed. If a bitch mink is known to have a litter of kits in a particular area, it is sensible to sit out in a concealed position, down wind from the site, and pick them off as they emerge to feed or play. It may take a while, but most can be shot in a day, especially if the bitch has been killed first. When sitting out for mink I like to use the .22 rimfire so long as there is a safe backstop. If the kits are in a hole at the waterside this is not usually a problem. If the area you are hoping to shoot mink in is one that is heavily fished, then they can be quite easy to approach, as they will be well used to seeing people on the banks fishing and they can become quite daring in their approach. Often they will attack fish in angler's keep nets and will keep coming back even when forcibly chased off. In situations like these, all you have to do is sit at the waters edge as if you were fishing, with the shotgun across your knees, and wait for a mink to come within range. But please let any fishermen know in advance of your intended actions since you don't want to induce a heart attack on some poor unsuspecting soul.

7
Rodent

To me, rodent control means only one thing, rats! Love them or hate them they are one of nature's great survivors. If one animal was to be left alive on earth after a major catastrophe it would be the brown rat (*Rattus norvegicus*). The brown or common rat has only been recorded in Britain since the early eighteenth century. It is widely believed to have been introduced on ships coming from Russia and not Norway, as the Latin name would suggest. The brown rat is not to be confused with the black or ship rats (*Rattus rattus*), which were believed to have returned with the crusaders from south-east Asia. These are the rats responsible for the infamous outbreaks of plague which killed so many people during the middle ages. The plague was caused by a bacterium that was transmitted to man via the rat flea. The black rat is now confined to mainly port areas, although it is occasionally found in inland towns, especially those linked to ports by canals. In Britain the black rat lives only indoors and its range continues to contract (although it is occasionally found in sewers). The common brown rat is often found living near water on the banks of ditches and streams and in sewers. As such, open water does not hinder them as they are reasonably good swimmers. They dive well and can remain submerged for a considerable time, a thought to bear in mind next time you are sitting on the lavatory; they have been known to enter premises by swimming round the S-bend in the lavatory bowl. The reproductive properties of rats are legendary. They are able to breed at about 3 months of age when they become sexually mature, with a gestation period of between 24–28 days, they produce litters of eight or more young, which can raise the rat population alarmingly in a short space of time. There are many and varied methods of rodent control, some are purely sporting, while others are more serious pest-control solutions. The most frequently used and most effective form of control has to be the use of anticoagulant rodenticides. From the single rat under the shed, to large infestations on farms and in grain stores, the careful placing of bait stations will quickly bring the situation under control. Let's have an in-depth look at the various options available to the professional pest controller.

Anticoagulant Baits

Ninety percent of all professional rodent control will include the use of anticoagulant baits. It is simply the most effective and quickest way of removing large infestations of rodents. Either spooned down rat burrows or placed in proprietary bait boxes situated around perimeter walls, it will rapidly eradicate most rat colonies. It is available in a pellet, liquid, grain, or wax block formulations and will contain active ingredients such as Warfarin, Bromadiolone and Difenacoum (to name a few). Most rodenticides available today are known as second-generation baits (with Warfarin being first-generation bait). The second-generation baits were introduced to overcome the problems encountered by the discovery of Warfarin-resistant rats. They were not actually resistant as such. What was happening was that the rats were eating enough bait to make them feel ill, but not enough to kill them, so they avoided eating it again. With the modern formulations they multi-feed on the baits without feeling any ill effects, but by the time they start feeling unwell, they have eaten more than enough bait to prove fatal. If you are thinking of using anticoagulant baits, I strongly recommend you attend one of the many training courses available through the BPCA or similar training bodies, and learn how to use rodenticides safely and effectively. In untrained hands anticoagulants can be very dangerous. They can prove fatal if ingested by non-target species; cats, dogs, and yes, even children, have been known to eat carelessly placed rat baits with disastrous results. With the modern baits, because they are slower acting, they are safer in as much as, if bait has been eaten by a dog for example, there is time to get the animal to a vet where it can receive the appropriate treatment. The responsible pest controller should always have received the appropriate training for whatever pest control method one is required to use.

Spring Traps

Mk4 Fenn traps are the only serious traps the rural pest controller is likely to use in an outdoor situation. The break back traps, like giant mouse traps, are fine for the odd rat in the house or roof, but for what I call 'heavy duty' rat control you can't beat the Fenns. Yes, you can use Kania or the body grip type traps outdoors, but for sheer ease of use the Fenn is the most versatile. Set in wooden tunnels against the walls of farm buildings, they will account for plenty of rats during the winter months as they return from a summer out in the fields. Left on permanent sentry duty, a tunnel trap is an environmentally friendly alternative to the use of anticoagulants. Certainly if there were barn owls in the vicinity of the buildings, I would be looking to use traps to avoid the possibility of secondary poisoning. If you have a barn or stable that requires rat clearance, providing they are out of reach of animals or birds, Fenn traps can be set on favoured runs. Here I'm thinking of where rats regularly run along the tops of walls or across roof beams. If you staple the holding chains of a few Fenn traps to the underside of a roof beam and set the traps on top, you will catch rats as they run across the beams at night. I have purposely made a few tunnels specifically designed to fit on top of beams to form a cover for the traps. Rats also like to run along the tops of walls, especially at the rear of buildings and these too can be productive catching places. You can place the traps about 18 inches apart all along the top of the wall for its whole length if the rat numbers warrant it. Again, if you staple the traps to the wall plate and lean a couple of heavy tiles over the top to form makeshift tunnels, they will account for any rats that use the wall as a chosen route. As with all trapping the traps must be inspected regularly

Rats can be caught in Mk4 Fenn traps placed in tunnels alongside walls

Rat caught in Mk4 Fenn next to a barn wall (tunnel removed for picture)

at least once but preferably twice a day and any catches removed. If for any reason the traps can't be inspected for a few days, then leave them in position held by the safety catch, which has the advantage of letting the rats run over the traps without them springing. This will work to your advantage when you eventually return to re-set them; the rats should have no fear of the traps, having been stepping on them quite regularly, only the next time they tread on them, will be the last. When using traps for rats they must be scent free. By this I mean, they mustn't have been thrown into the back of the motor alongside the dog and the ferret box because you won't catch anything for ages if they have. It doesn't matter if the traps are not cleaned after a catch; indeed, they will work even better if they have the scent of a recent victim on them. With rat traps, they need to operate with a smooth action and be corrosion free. Rub them down with some rough emery paper every now and then and lubricate them with clear candle wax (not the scented type the wife has in the bathroom), or you can also use vegetable oil, which will work just as well. On no account should you lubricate with household or gun oil because the smell will linger for months if you do. If you have already committed such a heinous crime, boiling the traps in water containing a little soda will clean them up again. The rat is a very quick and nimble animal and, unless the trap snaps shut instantaneously, it can, and does, often avoid the crushing jaws of the trap. Once a rat has been missed by a springing trap it is very unlikely to make the same mistake again, it will avoid the trap like the plague (excuse the pun).

Other good locations to site Fenn traps are: in voids between partitioned walls if they are wide enough to accept a trap; under grain silos or pheasant feeders; under feeding or water troughs; and perhaps one of the best locations, where a rat enters a building through a hole. If a Fenn is placed just inside the entrance (covered by a tunnel), any rat leaving or entering the building via this point will have to cross the trigger plate, with fatal consequences. Tunnel traps set along water courses, ditches and streams can be very productive catching places as rats regularly run the water margins to and from their favourite feeding places. Placed on well-used runs, tunnel traps, once they have become accepted, will account for many foraging rats as well as the odd stoat or two, which also use the rat runs when hunting them, the rat being one of their favourite foods. As mentioned in the chapter on mink control, if water voles frequent the areas you intend to trap, then you **must not** use spring traps for obvious reasons. We have touched on the subject before when discussing mink but, make sure you wear protective gloves when working in rat-infested locations and certainly when handling dead rats. Weil's disease is not very common but it can be fatal and extreme care must be taken at all times, especially when working in damp locations. Wet conditions are needed for the bacterium to survive outside the rat's body. It is usually transmitted to humans by contact with contaminated water or soil, or direct contact with rats. The organism enters the body through cuts and abrasions, or through the membranes in the nose or mouth. Weil's disease can range in severity from mild flu-like symptoms to renal failure and death. I know a gamekeeper who came very close to dying after contracting Weil's disease from a ferreting expedition in a rat-infested ditch. It was only the intervention of his fellow keeper who, after visiting him at home, and finding him delirious, semi-conscious and yellow, managed to get him admitted to hospital where they were able to save his life. As it was, he spent several weeks in intensive care and many more months recovering at home, so always be aware of the hazards when working with rats, and take sensible precautions.

Cage Trapping

If mink and squirrels are the easiest animals to catch in cage traps, then the rat is certainly the most difficult. With young immature rats the multi-catch type cages will catch them with ease; they have yet to learn the trait of new object reaction (neophobia). With mature adults though, it's a different story entirely. Any new object suddenly appearing in its territory is immediately treated with the utmost suspicion. I personally, rarely use cages for rodent control, preferring instead the Fenn trap or anticoagulant baits. I have placed a cage in a cellar baited with cooked sausage for over a week, knowing full well a rat was in there somewhere and unable to escape. The rat never entered the cage and the bait was left untouched. The client unbelievably, didn't want the rat poisoned or killed, she wanted it catching and releasing somewhere in the countryside (yeah right). In the end I persuaded her to let me humanely dispatch it, whilst she left the house unable to bear the thought of its demise. As soon as she'd gone, I moved a few boxes in the cellar, blocking off any escape routes in the process, and proceeded to hunt him down. Armed with my favourite ratting tool (golf club with the head missing), I gingerly picked my way through the debris. He didn't take a lot of finding, and made a dash for the door as I lifted a box in the corner; finding the exit blocked he made back for the corner where he came from. He never made it; a swift blow from the ratting tool rendered him lifeless on the floor. I wanted to do this a week earlier but you have to try and please all of the people all of the time in this business, so I had to try catching it first as she wanted. If there is no other means of escape, a cage trap will work if it is positioned outside a hole a rat has bolted into, although you may have a very long wait. Eventually the rat will have to leave his sanctuary to find food and water, and will have to make a dash for freedom into the cage. The cage entrance must be pushed tight against the hole and the cage weighted down with a heavy brick otherwise the rat will squeeze around the side and escape. So yes, cage traps can and do work sometimes, but there are a lot better and quicker methods available so I wouldn't recommend cages as a favoured option.

Shooting

Rats can be shot quite easily if you are prepared to sit quietly and wait for them to appear from their harbourages. I have sat at a window many times and shot rats as they've travelled backwards and forwards to garden bird tables and hanging feeders to raid the food supplies. If there is an abundant food source it's sometimes difficult to entice the rats to take rodent bait however palatable it may be, so unless they can be tempted into a tunnel trap, shooting is the quickest option. I use the .177 air rifle in these situations; it is plenty powerful enough to dispatch a rat at reasonable distances, although most shots are taken at short range. If the garden has a security light that illuminates the target area so much the better, as the rats feed with more confidence after dark and are also less worried about being seen. If you leave the bodies where they fall, and remain still, other rats will continue to emerge, completely unconcerned that one of their brethren is lying dead on the floor. On occasions I have witnessed rats trying to drag the lifeless corpses back to their burrows, presumably to feed on them later. Cannibalism is quite common among rats. If the rats are raiding the hanging bird feeders, you will very often shoot them climbing through the branches trying to reach them; I have shot many rats actually swinging on the feeders trying to get at the food. Often a combination of methods is required to achieve a successful outcome; if the rats have been raiding the bird feeders for some time, removing the food source and replacing it with anticoagulant baits will prove productive. One of the best methods to use here (and it works near pheasant feeders also), is to take a 2 ft × 2 ft paving slab and raise it up on four bricks, one in each corner. Underneath the centre of the slab, dig a shallow depression just deep enough to take the cut-off bottom of a plastic container (a washed out milk carton will do), and fill it with rodent bait. Further bricks can be placed under the edges of the slab to leave a couple of access points each side. The rats will quickly start to use the feeding chamber on a regular basis. When the bait stops disappearing you will know the job is done. One of the strangest jobs I was called to look at was to a very eccentric old lady who had rats regularly climbing a little Magnolia tree to get at the hanging bird feeders. The tree was situated just outside her conservatory window, where she had spent hours watching the 'baby squirrels' climbing the tree to get at the food. Her son, on a rare visit home, alarmed at discovering his mother was actively encouraging the local rodent population to expand, pleaded with me to rid the garden of the unwanted visitors. We hatched a plan whereby he would take his mother out for the day, and I would sit in a comfy armchair in the conservatory, with the double doors open, armed with the air rifle. There was a very high stone wall to the rear of the property, so I had a safe backstop if any pellets missed the target. During the course of a morning I shot 14 rats of varying sizes from the boughs of the tree, I can honestly say it was one of the best morning's sports I'd had for a long while.

Shooting rats around farm buildings after dark can also prove a profitable exercise, especially around cattle feeding troughs or grain storage areas. Sitting out in a concealed vantage point with the air rifle fitted with a lamp and red filter, will account for quite a number before the rats get wise to the strategy? When they do, it's best to leave it for a week or two, and then go back and have another session at them. I often sit in the 4×4 with the rifle resting on the wing mirror, it's warm and comfortable and, you can even have the radio playing provided it's on a very low volume.

Ratting with Terriers

As a sporting spectacle, watching a pack of working terriers decimate the local rat population takes some beating. Bolted from their lairs by the smoke from a chain saw engine, the dogs make short work of any rats that decide to vacate the premises. Usually carried out after the game covers have been flailed, it is a very efficient way of dealing with the large infestations of rats that have been allowed to multiply unhindered throughout the shooting season. Rats absolutely love setting up home amongst the maize cover crops where there is an abundant supply of food both from the crop and the regularly replenished bird feeders. Little wonder then that very large numbers of rats can be killed by a carefully administered operation. I must admit I do very little terrier work myself; I am first and foremost a lurcher man who specialises in rabbit control. That said, I have been on a couple of rat hunts to see how it was done so to speak and, I must say, I have been mightily impressed by the results. I have also been on hand to witness terriers at work when a barn has been emptied of straw bales, watching them dash around the next bale to be lifted in eager anticipation of a bolting rat; you can't fail to get caught up in the frenzied excitement. I won't use my lurchers for ratting jobs for one simple reason. Although they could, and would, catch a rat in an instant, my dogs are very soft mouthed and return rabbits alive to hand. I don't want them to start crushing rabbits in order to prevent bites that have been inflicted previously by rats. The quickest way to instil a hard mouth in a lurcher is to take it ratting. I am a firm believer in horses for courses. Terriers are supremely well adapted for tackling rats, and lurchers are designed for catching rabbits, and in my opinion you should never mix the two. If you have the opportunity to use terriers on large rat infestations then you must do so, as this is often the most successful method of reducing large numbers of rodents in one visit.

Gassing

I have had reasonable results in the past gassing rats, although it would not be my first course of action. If the rat burrows are open and accessible and the soil is clay based, as opposed to sand, then it can be successful, but there are many places where the use of gas is prohibited. The locations where aluminium phosphide cannot be used are those places most frequently used by rats such as watercourses, ditches, and around buildings, so for the pest controller it has very limited use. That said it is pretty effective for dealing with the odd isolated burrow system, those found in the middle of game cover for example. As always, you need proper training and instruction before you can use or purchase gassing compounds, so if you are thinking of gassing rats then enrol on a recognised training course and learn to do the job safely.

Ferreting

The use of ferrets to bolt rats has been around for a very long time although you don't tend to hear much about it nowadays. We used to do it as kids, using the legs off pairs of ladies tights, cut to the appropriate length; in one end we tied a knot, and the other was tied over an old baked bean tin that had both ends removed. You could stuff the open end of the tin into a rat hole, and any rats that bolted got caught in the tights where they were dealt a swift blow with a stick. We used to catch a good many rats using this method, especially around the muck heaps on the farms. I still use a ferret to shift the odd rat today, mainly from under a shed, in a cellar or a small barn. They are handy to use where a terrier would be unable to get, such as piles of old farmyard junk for example. Some serious ratters of the old school used to knit purse nets specifically designed to catch rats; with a half inch mesh, they would purse exactly the same as a rabbit net when a bolting rat hit them. The rats had to be dispatched pretty quickly when in the nets though, otherwise the razor sharp teeth would soon render the nets unusable. Very small Jill ferrets were the serious ratters preferred option, and if you had one that relished the job you were very lucky indeed. Ferrets often used specifically for ratting, would soon grow weary of the fierce fights involved (especially if tackling a doe rat defending her young) and would often refuse to enter after rats again. Let's face it, if you were put up against Mike Tyson six days a week, you'd be reluctant to face him again. Ferrets can take some serious punishment at times from cornered rats, and rabbits too for that matter, so it's sometimes understandable that they eventually tire of the ratting game completely. I am the opposite of the old time ratters; I prefer to use a very large polecat Hob on my ratting jobs. OK he can't get into some of the tiny spaces rats sometimes frequent, but if any rat is foolish enough to turn around and take him on, it will pay the ultimate price for its stupidity. That's a little unfair on the rat, because the last thing it is, is stupid. The rat is a very clever adversary most of the time, but like most animals, can act stupidly in times of sheer panic. I think rather too much of my gentle little rabbiting Jill's to put them up against rats. Indeed I would never do so, but the big polecat Hob is a different kettle of fish altogether. Though friendly to handle, he can turn into a psychopathic killer where rats and rabbits are concerned. Ferrets are handy at times on certain ratting jobs if only to locate and bolt the odd buck rat that has taken up residence in the garden shed. They can save an awful lot of work shifting boxes of junk and then having to put it all back again. To most people the rat is public enemy No.1. When it comes to the fear factor, normal rational people, for some reason, go into blind panic at the thought of a rat at large in the home. They will tell you tales of rats jumping up at the throat, of rats nibbling baby's ears and toes while they sleep, rats as big as cats (every job I go on has the biggest rat I've ever seen) and all sorts of old wives tales surrounding the rat. It must all stem from some primeval fear we've had bred into us from our cave-dwelling ancestors. The truth is this, rats don't jump up to attack the throat; what they do when cornered is jump up and try to get past you to escape. I have had this happen on a couple of occasions and pretty unnerving it is too. Once, when on my hands and knees searching for poisoned rats under the floor of a kebab shop, two rats very much alive, decided to leap past my shoulders and over my back to escape. I was injured during the encounter, though not by the rats. Startled, I banged my follicle-challenged head on one of the wooden floor joists, raising a very large lump in the process (on my head, not the joist). The second time it happened, I was trying to locate the entry point a rat was using to get into a house, when one jumped over my shoulder as I crouched down to move the washing machine. This time I was uninjured apart from my hearing that was seriously impaired by the demented shrill screaming of the young lady client. As it turned out, the rat was gaining entry through the cat flap (quite a common occurrence) and was helping itself to the cat food on a nightly basis.

But back to the ferreting of rats. Should you decide to use the ferret in some rat burrows on a farm for instance, make sure that the holes are not surrounding a manhole cover or anywhere near a land drainage system. Very often the rats use the drainage pipes to traverse the farmyard, and you can end up with the ferret reappearing hundreds of yards away down a ditch side somewhere. Also on the subject of drainage pipes, should your ferret suddenly appear in a yard full of cattle for instance, unless you are very fortunate, you will have one dead ferret to take home. I won't put ferrets down pipes in a farm situation, even if I can see both ends of the pipe. There is very often a junction somewhere along the length of the drainage system that you are not aware of. All pest controllers should have this motto imbedded in their brains: 'If in doubt, don't do it'. It is always better to be safe than sorry.

8
Fox

The fox is the one animal I get very few calls to control. I live next to a large country estate that runs a very productive partridge and pheasant shoot and as such, the fox is not tolerated on its land. The two keepers keep a constant vigil for any signs of fox activity both in the grounds of the estate and the outlying districts, most of which belongs to the estate anyway and, as such, the fox numbers are strictly regulated. Very often a client that has a supposed problem with a fox can be persuaded that to do nothing is the best option. Here I'm thinking of the fox with a litter of cubs under a garden shed or summer house, which will resolve itself in time when the cubs and vixen move on, or in other circumstances, like foxes using the garden as a thoroughfare, where it may only need the access point blocking off to achieve success. In most cases, customers will, after a little reassurance, learn to tolerate a fox using their property and may even begin to enjoy watching them as they go about their daily (and nightly) routines performing their playful antics. Most of my fox work involves helping out on small private shooting syndicates that have been losing pheasants or partridges as a result of fox predation. Here the solution is pretty straight forward; if I know from which direction he approaches the release pens, I can either sit out at night with the rifle and wait for him to appear or, if there is a discernible track or run, intercept him with a couple of discreetly placed snares. My preferred option is always to sit out with the rifle and lamp after dark and try to squeak the fox into a safe shooting range. Sometimes, due to the logistics of the site, the use of the rifle may not be practicable and a shotgun with a heavy load may be needed. As in all pest control, every job you attend is different and has to be judged on its particular merits. Some solutions may work better than others in certain situations and this is where your experience plays a major role. You should be able to survey any potential contract and instantly have a solution or solutions (there may be more than one) spring to mind, but if you find yourself scratching your head with a bemused expression on your face, then maybe you are not the man for the job. Nothing is more off putting to a potential client than someone who appears not to know what to do. Take your time when checking the site with a client. Stop here and there periodically, and examine tracks and entrance points (even if there aren't any), say very little, and wear a knowledgeable expression on your face. All this will instil confidence in the prospective customer and may make the difference as to whether you get the job or not. Don't go to the other extreme though and make out you know everything. This is equally as off putting to a client, and he'll probably know as much as you do anyway. OK, so you've got the job, now let's look at the options available to us in more detail.

Cage Trappping

The only legal trapping device (apart from a snare) for fox control is a cage trap. There is no other trap available that will catch and hold a fox and at the same time comply with the latest trapping legislation. I will occasionally use a cage trap where discretion is needed, a garden for instance, or where snaring and shooting is not practicable or possible. Foxes, like rats, are very clever animals and as such, are not easily enticed into a cage however tempting the bait may be. Young fox cubs though can be foolish in the extreme, and will sometimes enter even an un-baited cage out of pure curiosity. I have in the past, when called to remove cubs from a garden environment, left a cage in the set position, with no bait in place, to let them become accustomed to the trap, and on several occasions have caught quite well grown cubs in the cage. Now this may be that the cage smelt of a previous captive or some left over bait residue, but nevertheless, it demonstrates that an animal's

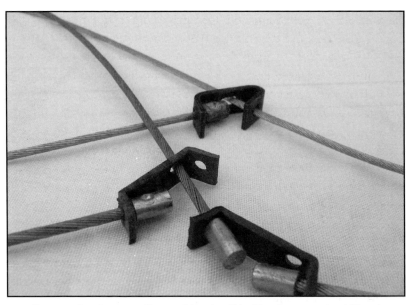

Free-running fox snare (bottom) legal, self-locking fox snare (top) illegal

Standard drop-door fox cage trap

natural inquisitiveness can often be used to our advantage. The positioning of a fox cage trap is vitally important; your natural instinct will be to set it where the fox has an earth or where a vixen has cubs under a shed. Although you will sometimes catch foxes here, there are a lot better places to set them that will increase you chances of success. Look at it logically; if you were a fox would you expect to find a damn great cage with a rabbit hanging in it, right outside the front door? I don't think so. When dealing with a fox in a garden, remember that he is basically a scavenger and as such, has probably been raiding every bin bag in the local area in his regular search for food. He will have been feeding on old chicken carcases, meat bones, half empty pet food tins and the remains of Friday night's Indian takeaway. Placing a cage where he **expects** to find food then, is the sensible option. When dealing with bin-bag foxes (as I like to call them) place the cage where the bags or bins are normally stored, cover and surround it with refuse-filled bags if necessary, and place a split-open bin bag with the contents falling out in the back of the cage using an old chicken carcase as bait. You can if you like, scatter a few tin foil takeaway containers, pet-food tins, old spare-rib bones or such like in the back of the cage as well, just to add a touch of authenticity to the picture. Now we have presented the fox with a dilemma, he may be suspicious of the cage at first, but as he has been feeding safely and undisturbed at this location for months, his natural instinct tells him that it is probably alright to enter the trap. If trying to catch a vixen and her cubs, I'll always try to place the cage where she has been regularly travelling in search of food. It may well be she's been foraging some distance away from the earth; if so, place the cage where she will pass it on her way out to feed. Disguising the cage with a lot of vegetation often helps when trapping rural foxes. If you leave the back end of the cage vegetation free, it will convey the image to the fox that it could walk straight through the cage if it wanted to. I have a cage that I use for catching cubs which has a clear Perspex sliding panel at the back of the trap. This works a treat especially if I can position it between a couple of walls so that the cubs think they can pass straight through it.

The best baits to tempt foxes into your cages depend very much on the location you intend setting them. As already mentioned, you must present the type of bait a fox would expect to find. In an urban garden, for instance, the fox is unlikely to have ever encountered a wild rabbit, so using one as bait would seem completely alien to it. With the rural fox, the rabbit is probably one of the best baits you could use, for the simple reason it forms the bulk of his staple diet. Try and visualise what type of food the fox has been feeding on and bait the cage accordingly. A close inspection outside a fox earth will usually offer up a few clues; chicken wings or pheasant feathers, rabbit bones and fur scattered about, will tell you exactly what you need to use as bait.

If trying to catch a poultry-killing fox, a cage placed just inside the entry point baited with the remains of one of his victims will tempt him into the trap; it may take a few nights, but he will come back. When I know there is a fox-trapping job in the pipeline I save all the guts from my rabbit paunches and keep them bagged up in the freezer (don't blame me for your wife's reaction). Once I have found where I will position the cage, I dig a shallow pit underneath the cage at the baited end and fill it with the thawed out rabbit guts. I then hang a fresh rabbit up at the back of the cage with the entrails removed; making sure that the fox cannot reach it and pull it through the wire.

Bedding the bottom of the cage down into the ground often helps; the fox is then entering the trap walking on earth instead of metal, which would immediately arouse suspicion. Don't clean the cages after they have caught. Fox cages need to be as smelly and obnoxious as possible; you may not like it, but the fox most certainly will. As with all trapping the cages need checking daily and the occupant's humanely dispatched or transferred to another location and released, as is often a client's wishes. The best weapon to destroy the fox in a cage is the .22 rimfire. A clean shot to the head at the base of the skull (where the vertebrae joins the head) and death will be instantaneous. Having to destroy animals is not the nicest feeling in the world, but if you work in pest control it is something you have to come to terms with. If you can ensure a swift and humane end to any creature's life, then that's the best you can hope for, and if you can't, then I'm afraid you shouldn't be involved with pest control.

Snaring

At the time of writing it is still legal to use free-running snares for fox control (not the self-locking type which are illegal). A new code of conduct was launched by DEFRA in October 2005 which sets out best practice when snaring foxes. The new code, entitled the **Defra code of practice on the use of snares in fox and rabbit control** was the result of a complete review of the use of snares in England and Wales. One of the main points raised was the fitting of a stop on a fox snare wire so it became a restraining rather than killing device. The code also states that snares must not be set in sites cluttered by obstacles, such as saplings, hedges, walls, fences or gates which could increase the risk of injury as a result of the snares becoming entangled. Anyone considering using snares for fox control should obtain a copy of the code and adhere to the guidelines contained within it. Snares need preparing and weathering the same as in rabbit control, before they are ready for use. You can boil them first in a pan containing a little soda (or automatic washing powder) for an hour, and then boil them for a further hour in a pan containing pieces of oak bark, walnut shell husks, or tea bags. They then want leaving in the cooling water for about 24 hours to let them stain.

After the boiling process, a little re-waxing of the snares is beneficial; you can use clear candle wax or paraffin wax neither of which has much odour. On no account should you use household oil, gun oil, or aerosol lubricant. At this stage it is a good idea to fix a 'permanent stop' to the snare wire, fixing it approximately 9 inches (23 cm) from the eye of the snare. It's now a good idea to hang the wires outside to weather or until you are ready to use them.

Unlike rabbit snaring there is seldom a fox 'run' as such; he will use existing pathways, tramlines in growing crops, sheep tracks, wheel ruts and dry ditches, indeed anything that affords him a discreet approach to his desired destination will be used. Only rarely will he be viewed out in an open field, preferring instead to hug the fence margins. If he is spotted in open ground, he is either hunting, responding to a call, or has been forced into the open to take the quickest route home. Fox snares need a tealer to hold them from the ground as in rabbit snaring, but unlike a rabbit tealer, the one for fox snares wants fashioning out of a length of stiff wire; set up beside the run they become almost invisible when standing in vegetation. The noose of a fox snare wants to be roughly 7 inches (15 cm) deep and 9 inches (23 cm) across, a sort of drooping pear shape, and needs positioning with the bottom of the loop at least 8 inches (20 cm) from the ground, which can be increased to 12 inches (30 cm) in open ground. If there is a possibility of deer in the vicinity, a thin hazel wand can be loosely bent over the set snare; this does one of two things, first it will encourage the deer to jump the obstacle, and second it will get the fox to duck his head to go under it. I prefer to set my fox wires a little on the high side; this dramatically reduces the risk of accidentally catching a badger (he will easily brush under it if he happens to use the same trail as the fox). On the subject of badgers, snares must not be set on or near a badger sett, or indeed on any runs that radiate from it. If badgers frequent the location you intend snaring, my advice would be to find an alternative method of dealing with your fox. Snares should be firmly anchored down using a heavy stake hammered well into the ground, and if you keep the stake low to the ground you minimise the risk of fatal entanglement by a captive fox. If you set fox snares on a regular basis, sooner or later you may capture a non-target species, such as a deer, badger, cat or dog. If you are checking the snares twice a day (and you shouldn't be setting snares if you don't) most non-target species can be released with relatively little damage. Beware though if you have a trapped badger or cat to deal with as they will inflict a very savage bite to anything they can get near to. The best way to release a trapped animal caught by the neck is to take a garden fork and slip the tines over the wire at the anchor point, and then slide the fork up to the animals head. Then push the fork into the ground without using your foot (if you don't want to get bitten) which pins the animal by the neck. A blunt stick or hook can then be slipped under the wire enabling it to be lifted from the neck and then to be cut with wire cutters. Fortunately it is a rare occurrence if you catch a non–target animal, especially if you take all appropriate measures to avoid it.

I would only recommend the setting of snares for foxes if you are really experienced in rural pest control. You must be able to differentiate between the tracks of a badger and the path of a fox, between the path of a sheep and the tracks made by deer. Some footpaths are barely discernible through growing crops, but somebody might be using it to walk the dog. Remember the pest controller's motto? 'If in doubt, don't do it'.

Shooting

The most efficient form of fox control has to be shooting, whether it is on organised fox drives, or calling up to the gun and lamp at night, it is simply the best and quickest way to deal with a fox. Fox drives can be productive where there are large areas of gorse or scrub; foxes tend to lie up in these areas during the day, usually in a warm sunny spot where they remain undisturbed until their night time forays in search of food. Fox drives need to be well organised and thought out. Guns must be positioned at every possible escape route to intercept the fleeing fox, because if you leave one unattended this will be the one he will use to make good his escape. Since the ill-conceived Hunting Bill was introduced, you may only use two dogs to flush a fox to waiting guns, which in large areas of woodland is pretty ineffective. You can make up for the lack of dogs though by using lots of beaters walking in close formation; if they are kept in as straight a line as possible, not many foxes will be missed. A shotgun using a heavy cartridge load is the weapon of choice for fox drives; they are perfectly suited for the snap shooting that is often required to stop a bolting fox. When positioning guns on fox drives, as well as the waiting guns up in front, it is beneficial to have two walking guns on the flanks, following slightly back from the line of beaters. Foxes, if sensing trouble ahead will often double back or slip out of the sides of a beating line, so having a couple of guns bringing up the rear can often pay dividends.

The most popular method of shooting foxes is with the rifle and lamp at night. Called into range by squeaking, unless it has been fooled before, the fox will come running to investigate the strange sound thinking a free meal is in the offing. A fox will run and stop several times as it approaches the waiting gun, and providing you are downwind of the fox, will eventually come near enough to provide an easy target. How you choose to make the call is a matter of personal choice. Most call through pursed lips, as indeed I do, some use a shop-bought fox call, others use a small piece of moist polystyrene rubbed on the windscreen to fool the fox (I can't stand the noise myself). Fox calling is particularly effective when dealing with young cubs and I have had them come running towards me at such speed that I couldn't safely take an effective shot. On the other hand, if a fox has been called up on a previous occasion and the shot missed, it is very unlikely you will deceive him again. As with all rifle shooting you must be familiar with the ground you are shooting over, and be sure you have a safe backstop at all times. When you are using powerful rifles, such as the .222 or .243, safety is paramount. Always make sure of the identity of the target before pulling the trigger, and **never** ever fire at only a pair of eyes as there have been some tragic accidents of late by somebody doing just that. Foxes can be called into range during the daytime provided you are well concealed and working into the wind. Often more than one fox at a time will respond to the call. Sometimes I have been calling one fox when another has appeared to the side or behind my position; choose the safest shot in this instance for you will only take one.

I have only briefly touched on the subject of fox control for the simple reason that most pest controllers will seldom get asked to deal with them. As I have stated previously, fox control is best left to the experienced professional and shouldn't be undertaken by amateurs. If you are unsure of your abilities to control foxes then don't undertake this kind of work; as always stick to what you know. You are not obliged to control any species of animal; if asked to do so, you can always decline the offer of fox control or any other type of work if you so wish.

9

It Shouldn't Happen to a Pest Controller

In the preceding chapters we have looked at some pretty serious aspects of rural pest control. As a way of some light relief, I thought you may like to share some of the moments during my illustrious career when things didn't always go quite to plan. I will start a long time ago, in the days of my youth, years before I became a full-time pest controller. There used to be some old brick workings behind where I lived as a child and these had stood empty and abandoned for many years. The gorse strewn sandy wastelands were a haven for wildlife. Rabbits, hares, partridge and pheasant were abundant, linnets and bullfinches nested in the many hawthorn bushes that were dotted over the site, and the song of the skylark could be heard in the skies overhead. Traversing the site were many pathways and tracks and these were used regularly by the local rabbit population. In my young and furtive imagination, I reckoned if I dug a hole about four feet deep and the same across, and covered it with twigs and dried grasses, sooner or later a rabbit would be bound to fall in it. Several of these pits were dug (I think I had been watching too many Tarzan films) and were duly covered over. I didn't have long to wait before my first catch, only it was not the rabbit as I expected. No, it was an elderly gentleman out walking his dog, who suddenly disappeared from view as I watched from a safe distance. I didn't stop to see if he was trapped or injured, but mortified, I turned on my heels and headed for home as fast as my legs would carry me, hoping and praying that nobody had seen me. I spent a few uncomfortable nights re-enacting the incident in my mind, unable to sleep in case I'd killed the old chap, but nothing came of it and it was soon forgotten.

During the many years I have been involved in pest control some weird and wonderful events have occurred whilst carrying out my daily tasks. Some are funny, some painful, and some are downright dangerous, but most are interesting to say the least. Being out and about at all hours, many strange sights and sounds have been encountered; many times at night I have been halted in my tracks by some fearsome apparition, only to find it's a bullock or horse leaning over a fence. At night, many farm animals take on very strange appearances, with luminous eyes, and sudden human-like coughs that can startle even the most wary of travellers. Once, on a very dark night whilst out lamping, I glimpsed at the very limit of the lamps range, a figure watching me from behind a hawthorn hedge. As I was using the .22 rimfire for shooting rabbits, I was naturally concerned that the person was aware of the potential danger. Not wishing to blind the person with the lamp, I switched it off and made towards him to have a few words. On getting closer I switched the lamp back on and immediately fell about laughing. There in front of me was an old ewe standing up on her hind legs trying to feed on the hawthorn buds. Feeling rather sheepish (if you will excuse the pun) I crept back off into the darkness to continue the night's work.

It's not only at night that animals can cause panic and alarm. As a lad, I once sat quietly fishing on the riverbank when around a bend in the river came a hippopotamus. I dropped my fishing rod and retreated at a rate of knots to the top of the bank. I was genuinely terrified for a few moments until I realized that the head complete with revolving piggy ears and flaring nostrils was a black bullock that had somehow managed to fall into the river. The bedraggled creature appeared slate grey in colour due to the covering of water, and it was an exact replica of the real thing. With just its head visible the poor creature meandered through the lily pads and continued on its journey downstream, where I watched it scramble out to regain dry land.

One of the disadvantages of being a full-time pest controller is that from time to time you feel the wrath of some of the species you are called on to deal with. I treat in excess of 300 wasp nests each season and getting stung once or twice is an occupational hazard and I accept that as part of the job. With the exception of a mink and a badger, I think I must have been bitten, stung, or kicked by nearly every creature there is in this country. Ferret bites, although painful at the time, are not too bad; rat bites are far worse and often require medical intervention. The friendliest house cat can turn into a spitting snarling demon if accidentally caught in a rabbit or fox snare, and cat bites, especially if it is a feral cat, invariably fester and turn septic. Even the wearing of protective gloves can sometimes not be enough to prevent painful injuries. Once while trying to remove a trapped grey squirrel from behind the down pipe of a gutter to a school roof, I was badly bitten on the finger even though I was wearing thick leather gloves; the squirrel's razor sharp teeth had cut straight through the glove. On the subject of painful injuries, if anyone has experienced the pain of a trap going off on their fingers (and I don't mean mouse or mole traps) they will know exactly what I am talking about.

Twice I have been unlucky enough to have had a Mk6 Fenn trap go off on my fingers, and I wince even now just writing about it. The first time I was clearing the debris from a rabbit hole prior to re-setting the trap. The Fenn was balancing on top of the hole in the set position with the safety catch engaged when it started to topple into the hole. To this day I do not know what possessed me, but I instinctively grabbed at the trap to stop it falling. At the precise moment my fingers touched the trigger plate, I saw the safety catch fall off. Now anyone who knows me understands that I swear a little, but I let forth a torrent of expletives like you'd never heard before. The trap sliced clean through a gold ring I had on my finger, and I believe it was only this that prevented my fingers from being broken. As it was my fingers were bleeding and badly bruised. Now this is not the end of the tale because I now had a Mk6 Fenn securely fastened to my four fingers that needed two free hands to open it. It was a freezing November day,

I was miles from anywhere with numb and bleeding fingers, and I had visions of walking back to the vehicle and driving home still attached to the trap. Luckily I had my mole dibber bar with me and I was able to force the jaws apart just enough to extract my painfully swollen fingers.

The second time wasn't quite so bad but was equally as painful. The trap was positioned within the rabbit hole and I had just flicked off the safety catch. Again, I don't know what made me do it but I wasn't happy with the soil covering the trigger plate and foolishly reached over the pan to pull some soil from the back of the hole. Wrong! This time the trap grabbed me across the hand from thumb to little finger, which meant the jaws were held quite wide apart, and I was able to extract my hand with very little effort.

Anyone who does a lot of rabbit-cage trapping will be well used to finding non-target species caught in the cages. Foxes and badgers are often caught in cages while trying to get at the captive rabbits and badgers can do immense damage to a cage in its efforts to escape. A trapped badger can sometimes be a nightmare to get out of a cage. I once had a huge badger stuck in a rabbit-cage trap; he was that big and fat (especially after devouring the rabbit) that he was wedged in tight. His head was at the blocked end of the cage and I had to stand the cage on end and physically push his backside down into the cage so I could open the door. I could then push a stick through the wire to hold the door open, and with my mole dibber bar through the cage at the head end, I was able to shake him out on the other side of the rabbit fencing where he trundled off quite unperturbed.

Moving on to rabbits, don't let anyone tell you rabbits don't bite, because they most certainly do. Whilst checking a snare line early one morning, I came to a snare that had the tethering cord chewed through; it had obviously had a rabbit in it due to the flattened grass around the peg. About four snares further along, I removed a rabbit from a wire that had the missing snare around his neck; he was one unlucky rabbit. I once caught two rabbits in the same snare, one had been noosed around the neck in the normal fashion, and the other had somehow managed to get tangled in the wire as it came to investigate the trapped rabbit; the wire was twisted and looped around its middle. Anyone who has done any amount of long netting will testify to rabbits biting through the netting if left for any length of time trapped in the net, but that is nothing compared to the damage caused by an irate cat if it becomes entangled in one. I once caught the farm cat in a long net when ferreting around some hay bales stacked in a field. I was bitten and clawed mercilessly before finally cutting it loose.

I don't know what it is with pest controllers, whether it's because we creep quietly around at night, or sit silently at some remote vantage point, but it seems almost obligatory that local courting couples are often encountered. I have lost count of the number of times I have found couples, doing what courting couples do, when out and about at night. Twice I have clambered through a gap in the hedge at night and landed, yes landed, on top of a courting couple (not the same couple I might add). The resulting screaming and shouting of all three parties' leaves a lasting impression on ones mind. You are not safe in the daytime either. One hot summer's day I was walking quietly along the hedgerow stalking outlying rabbits; I had my two whippet lurchers with me at the time, both of whom were walking steadily by my side. From around a bend in the field up ahead of me I heard a crack, then another, like the sound from an un-moderated .22. Thinking someone was shooting rabbits with the rifle, I thought I would walk out into the field and let the person know I was there. The vision that greeted me when peering around that corner will remain with me for the rest of my life. Kneeling on all fours on the ground was a gentleman and standing above and behind him was an Amazonian, leather-clad dominatrix-type women, clasping a leather belt she was using to administer punishment to the man's rear end. I don't know who was the more embarrassed, me, or the couple who were now being closely examined by the two lurchers. Many times gate and long-netting sessions have been interrupted or curtailed by courting couples at night. It seems somewhat churlish to tap on a steamed up car window parked in a remote gateway and to ask the occupants if they would mind moving so you can set a gate net.

As I have stated previously, when treating wasp nests you expect to get stung occasionally; that is an occupational hazard. But if you think that treating an infestation of ants would be pretty uneventful, well think again! Once I was treating ants for a little old lady in a terraced house. The ants were entering her living room from the cavity behind the gas fire and marching across the carpet to the kitchen. The gas fire was switched off; naturally, it was the middle of summer after all. Lying on the floor in front of the fire I aimed the crawling insect killer aerosol can underneath the fire and began to spray. The tin was nearly empty when there came an almighty boom and flash from behind the fire. Well, my eyebrows had gone, my moustache was singed, and the ladies nice cream fireside rug was now a smouldering orange and black colour. In my haste to solve the ladies ant problem, I had forgotten about the pilot light that flickered constantly within the confines of the gas fire. She was very understanding about it though.

The pitfalls are many if you take up the pest-control profession. You never know what the next job will bring, and you have to be prepared for anything, and I mean anything, if you want to succeed in pest control. That said, it is one of the most rewarding and satisfying jobs you are ever likely to do, especially if you have a natural leaning towards the countryside and all things rural. So don't let these stories of misadventure stop you pursuing your dreams of becoming a pest controller. When fate strikes you a downward blow, just smile, carry on, and put it down to experience.

10

Conclusion

We have looked at the various rural species you are most likely to get called in to deal with as a professional pest controller. Some have been deliberately omitted like the corvids (crows and magpies), stoats and weasels, and feral cats to name a few. Most country-based pest controllers know how to make and use a Larsen trap (it isn't rocket science), and most of the mink or rat-trapping techniques can be adapted or modified to capture stoats and weasels. With feral cats, you can cage trap them if you wish, but you are then left to deal with them, or you can, as I often do, refer the client to the local cat protection league or similar body and let them deal with it. Also I haven't discussed insect control, wasps and ants for example. You are entering a whole new world of insecticides, legislation, risk assessments, etc. I could fill another book based entirely on the study and control of insects. If you want to treat insects as part of your daily workload, then enrol on one of the many training courses available and learn to apply the myriad of insecticidal treatments safely and effectively. In rural pest control you will be expected to solve any problem a potential client may throw at you. You will be asked to get rid of snakes (these are protected by the way), hedgehogs, bats (protected also), mice, and even frogs from a pond. I've been asked to remove all of these before now, so never be surprised as to what causes distress in certain people.

One aspect of pest control that is seldom mentioned is what happens if something goes wrong. If you are undertaking pest control work for members of the general public, you must have adequate public liability insurance cover (usually £5million). Many of the larger contract jobs you may be invited to tender for won't employ you unless you have this in place. We live in an age where compensation claims are commonplace; many companies exist solely for the purpose of extracting as much money from a liable source as possible. Even if you undertake pest control as a purely sporting exercise, it is advisable to belong to a shooting body such as the BASC who have their own insurance policies in place. Don't let all this put you off following your chosen profession though. If you undertake pest control as a livelihood it can be one of the most rewarding and satisfying jobs there is.

I am a firm believer that the hunting instinct is bred into certain people. I was born a hunter, don't ask me how or why, I just was, some primeval throwback perhaps, although none of my siblings were remotely interested in the countryside. As a young boy I can remember seeing a cock pheasant dart into the bottom of a hedgerow and thinking, how could I catch one of those? I would devise all sorts of devious contraptions in my mind's eye to capture one of these beautiful creatures, but who hasn't tried to capture beauty at some time or other? I was always happiest when I was on my own as a child, some remote spot as far away from other people as I could get was my idea of utopia. I still love to be on my own, whether it's standing in a spinney at dusk waiting to ambush the pigeon as they come to roost, or running out a long net after dark along the covert side on a wild and windy night; it's absolutely magical. It's my opinion that you only become truly alive and at one with nature if you can 'feel' the solitude and wonderment of being alone with the elements. Ask any wildfowler what he loves most about his sport, I guarantee he will say it's about being out on the estuary mud flats or salt marsh at dusk or first light, rather than the amount of birds he might shoot. It's the same with roost shooting pigeons, it's about being in a wood with the wind howling through the tree tops and the boughs creaking and groaning overhead that is the real attraction.

So there you have it, I have told you of what works for me in certain situations, not all the time you understand. Very often I fail the same as everybody else, we all do at one time or another. Some may not agree with all that is written here, preferring to use different methods entirely to those I've described, but I hope that I may have helped at least one or two people to have a better understanding of the complex nature of rural pest control. I'll leave you with this thought. There are more advantages than disadvantages to being a pest controller. You get to witness some wonderful sights at times; being out at first light you see some spectacular sunrises, and at dusk some sunsets are magical. I am privileged to see nature as it happens, and who can fail to be moved by the sight of a hare suckling two tennis ball sized leverets only hours old, or a mallard calling her newly hatched young down from the hollow in a willow tree to follow her to the water. Yes, it's very hard work at times being involved in pest control but I wouldn't swap my job with anyone.